The Basics of Geomorphology

The Basics of Geomorphology

Greg Powell

MURPHY & MOORE
www.murphy-moorepublishing.com

Published by Murphy & Moore Publishing,
1 Rockefeller Plaza,
New York City, NY 10020, USA
www.murphy-moorepublishing.com

The Basics of Geomorphology
Greg Powell

International Standard Book Number: 978-1-63987-522-1 (Hardback)

Cataloging-in-Publication Data

The basics of geomorphology / Greg Powell.
p. cm.
Includes bibliographical references and index.
ISBN 978-1-63987-522-1
1. Geomorphology. 2. Physical geography. 3. Landforms. I. Powell, Greg.
GB401.5 .F86 2022
551.41--dc23

TABLE OF CONTENTS

PREFACE

This book is a culmination of my many years of practice in this field. I attribute the success of this book to my support group. I would like to thank my parents who have showered me with unconditional love and support and my peers and professors for their constant guidance.

The scientific study aimed at understanding the origin and evolution of various landforms present on Earth is known as geomorphology. It includes the study of topographic and bathymetric features of the landforms. Climatic geomorphology focuses on climate's role in shaping the landforms and related processes. Biogeomorphology is an interdisciplinary field which studies the interactions between development of landforms and organisms. The study that focuses on interaction and relation of water and hydrologic processes with landforms is called hydrogeomorphology. There are diverse processes related to geomorphology. These include aeolian, biological, fluvial, glacial, hillslope, igneous, marine and tectonic processes. This book is a valuable compilation of topics, ranging from the basic to the most complex theories and principles in the field of geomorphology. Such selected concepts that redefine this field have been presented in it. For all those who are interested in geomorphology, this book can prove to be an essential guide.

The details of chapters are provided below for a progressive learning:

Chapter - Introduction

Geomorphology deals with the study of physical features of the Earth's surface and their relation with its geological and topological structures. A digital elevation model analyzes the data by estimating the values of elevation of Earth's ground surface. This is an introductory chapter which will briefly introduce about geomorphology.

Chapter - Branches of Geomorphology

Geomorphology can be categorized into a varied number of branches. Some of them are fluvial geomorphology, biogeomorphology, climatic geomorphology, hydrogeomorphology, applied geomorphology, etc. The topics elaborated in this chapter will help in gaining a better perspective of the different branches of geomorphology.

Chapter - Geomorphic Processes

Geomorphic processes refer to the physical and chemical processes that modify the surface of the Earth. It includes igneous process, aeolian process, fluvial process, hillslope process, marine process, etc. This chapter has been carefully written to provide an easy understanding of these geomorphic processes.

Chapter - Weathering and Erosion

The breaking down of rocks, soil and minerals when they come in contact with the Earth's atmosphere, water and organisms is defined as weathering. Erosion refers to the act of eroding the surface by wind, water and other natural agents. This chapter sheds light on weathering and erosion for providing an in-depth understanding of the subject.

Chapter - Landforms and Landscapes

A landform is a natural, geographical or artificial feature that appears on the Earth's solid surface. Landscape refers to the characteristic features of an area. Landscapes are divided into two categories - natural and human landscapes. This chapter closely examines landforms and landscapes to provide an extensive understanding of the subject.

Greg Powell

Introduction 1

- Geomorphology
- Geologic Structures
- Digital Elevation Models in Geomorphology

Geomorphology deals with the study of physical features of the Earth's surface and their relation with its geological and topological structures. A digital elevation model analyzes the data by estimating the values of elevation of Earth's ground surface. This is an introductory chapter which will briefly introduce about geomorphology.

Geomorphology

Geomorphology is the science of landforms, with an emphasis on their origin, evolution, form, and distribution across the physical landscape. Understanding geomorphology is therefore essential to understanding one of the most popular divisions of geography. Studying geomorphological processes provides significant insight into the formation of the various structures and features in landscapes worldwide, which can then be used as a background for studying many other aspects of physical geography.

Although the study of geomorphology has been around since ancient times, the first official geomorphologic model was proposed between 1884 and 1899 by the American geographer William Morris Davis. His geomorphic cycle model was inspired by theories of uniformitarianism and attempted to theorize the development of various landform features.

Davis's theories were important in launching the field of geomorphology and were innovative at the time, as a new way to explain physical landform features. Today, however, his model is not usually used, because the processes he described are not so systematic in the real world. It failed to take into account the processes observed in later geomorphic studies.

Since Davis's model, several alternative attempts have been made to explain landform processes. For example, Austrian geographer Walther Penck developed a model in the 1920s that looked at ratios of uplift and erosion. It did not take hold, though, because it could not explain all landform features.

Geomorphological Processes

Today, the study of geomorphology is broken down into the study of various geomorphological processes. Most of these processes are considered to be interconnected and are easily observed and measured with modern technology. The individual processes are considered to be either erosional, depositional, or both.

An erosional process involves the wearing down of the earth's surface by wind, water, and ice. A depositional process is the laying down of material that has been eroded by wind, water, and ice. There are several geomorphological classifications within erosional and depositional.

Fluvial

Fluvial geomorphological processes are related to rivers and streams. The flowing water found here is important in shaping the landscape in two ways. First, the power of the water moving across a landscape cuts and erodes its channel. As it does this, the river shapes its landscape by growing, meandering across the landscape, and sometimes merging with others to form a network of braided rivers. The paths rivers take depend on the topology of the area and the underlying geology or rock structure where it moves.

As the river carves its landscape, it also carries the sediment it erodes as it flows. This gives it more power to erode, as there is more friction in the moving water, but it also deposits this material when it floods or flows out of mountains onto an open plain, as in the case of an alluvial fan.

Mass Movement

The mass movement process, also sometimes called mass wasting, occurs when soil and rock move down a slope under the force of gravity. The movement of the material is called creeping, sliding, flowing, toppling, and falling. Each of these depends on the speed and composition of the material moving. This process is both erosional and depositional.

Glacial

Glaciers are one of the most significant agents of landscape change because of their massive size converts to power as they move across an area. They are erosional forces because their ice carves the ground beneath them and on the sides, which forms a U-shaped valley, as with a valley glacier. Glaciers are also depositional because their movement pushes rocks and other debris into new areas. The sediment created when glaciers grind down rocks is called glacial rock flour. As glaciers melt, they drop debris, which creates features like eskers and moraines.

Weathering

Weathering is an erosional process that involves the mechanical wearing down of rock by a plant's roots growing and pushing through it, ice expanding in its cracks, and abrasion from sediment pushed by wind and water, as well as the chemical break down of rock like limestone. Weathering can result in rock falls and unique eroded rock shapes like those in Arches National Park, Utah.

Geologic Structures

Geologic structures such as faults and folds are the architecture of the earth's crust. Geologic structures influence the shape of the landscape, determine the degree of landslide hazard, bring old rocks to the surface, bury young rocks, trap petroleum and natural gas, shift during earthquakes, and channel fluids that create economic deposits of metals such as gold and silver.

Folds, faults, and other geologic structures accommodate large forces such as the stress of tectonic plates jostling against each other, and smaller forces such as the stress of gravity pulling on a steep mountainside. An understanding of the structures that shape the earth's crust can help you see when and where the crust was subjected to pushing or pulling, terrane accretion or crustal rifting.

Physical Behavior of Rocks: Stress and Geologic Structures

Before exploring geologic structures, we need to look at how rocks respond to the forces that create the structures. Stress refers to the forces that cause rocks to deform. There are three basic types of stress that deform rocks:

- Compression (pushing together).
- Tension (pulling apart).
- Shear (twisting or rotating).

In response to stress, rocks will undergo some form of bending or breaking, or both. The bending or breaking of rock is called deformation or strain.

If rocks tend to break, they are said to be brittle. If a rock breaks, it is said to undergo brittle behavior. If rocks tend to bend without breaking, they are said to be ductile. If a rock bends but is able to return to its original position when the stress is released, it is said to undergo elastic behavior. If a rock bends and stays bent after stress is released, it is said to undergo plastic behavior.

A combination of elastic and brittle behavior causes earthquakes. Rocks get bent in an elastic fashion until they reach their limit, then they break in brittle fashion. The rocks on either side of a break act like rubber bands and snap back into their original shape. The snap is an earthquake and the break along which the rocks slide back to their original shape is a fault.

Earthquakes and faults occur in the shallow crust, where rocks are relatively cold and therefore brittle. In the deep crust and deeper, in the earth's mantle, rocks are very hot and subject to high pressure caused by the weight of the overlying rock. This heat and pressure causes deep crustal and mantle rocks to be ductile. In fact, rocks deep in the continental crust and upper mantle can be so hot and soft that they behave almost like a slow-moving liquid, even though they are actually solid. They "flow," or bend, or stretch, in a plastic manner, at a geological pace.

Now let us look at the specific types of geologic structures - the breaks and bends that deform rock in response to stress.

Folds

Ductile rocks behave plastically and commonly become folded in response to stress. Folding can happen in the shallow crust if the stress is slow and steady and gives the rock enough time to gradually bend. If the stress is applied too quickly, rocks in the shallow crust will behave as brittle solids and break. However, deeper in the crust, where the rocks are more ductile, folding happens more readily.

Anticlines and Synclines

The most basic types of folds are anticlines and synclines.

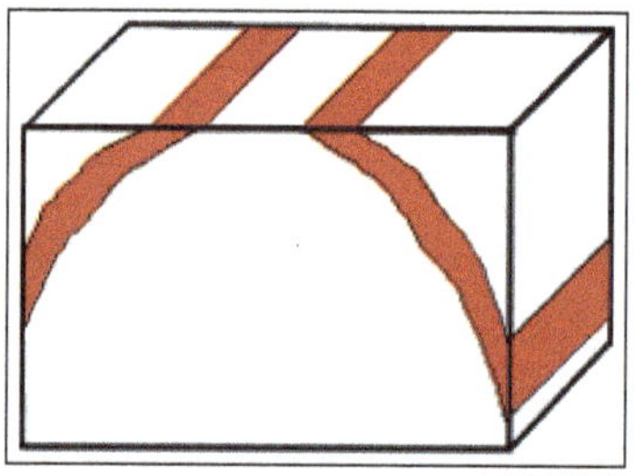

Anticline.

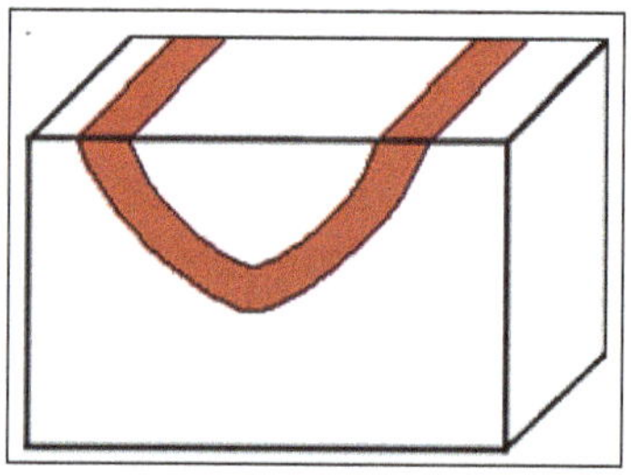

Syncline.

Anticlines are "up" folds; synclines are "down" folds. In box diagrams like these, the top of the box is the horizontal surface of the earth, the map view. The other two visible sides of the box are cross-sections, vertical slices through the crust. The colored layers represent layered geologic formations that were originally horizontal, such as sedimentary beds or lava flows. Use the block diagrams to visualize the three-dimensional shapes of the geologic structures. Keep in mind that erosion has stripped away the upper parts of these structures so that the map views reveal horizontal slices through the structures.

In map view, an anticline appears as parallel beds of the same rock type that dip away from the center of the fold. In an anticline, the oldest beds, the ones that were originally underneath the other beds, are at the center, along the axis of the fold. The axis is an imaginary line that marks the center of the fold.

In map view, a syncline appears as a set of parallel beds that dip toward the center. In a syncline the youngest beds, the ones that were originally on top of the rest of the beds, are at the center, along the axis of the fold.

Anticlines and synclines most commonly form in sections of the crust that are undergoing compression, places where the crust is being pushed together. Crustal compression is commonly the response to stress from more than one direction, which causes tilting as well as folding. In crust that has been tilted, folds "plunge" into the earth's surface.

Plunging Anticlines and Synclines

A plunging anticline or a plunging syncline is one that has its axis tilted from the horizontal so that the fold dips into the earth.

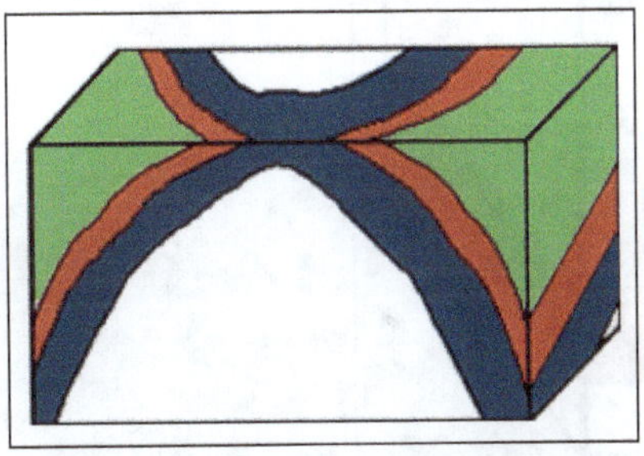

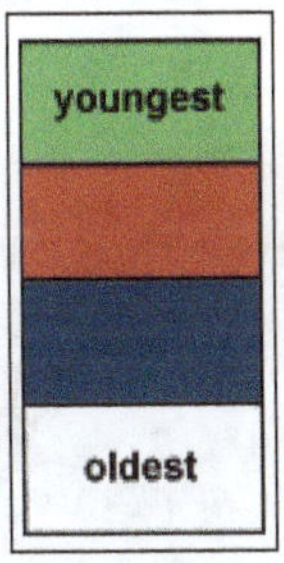

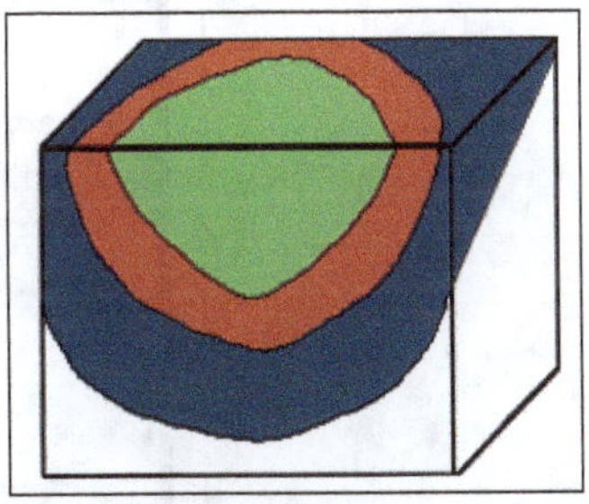

Plunging Anticline.

Plunging Syncline.

In map view, a plunging anticline makes a U-shaped or V-shaped pattern that points in the direction of plunge. (The plunge direction of the fold is the direction in which the axis of the fold dips downward into the earth.)

A cross-section across the axis of a plunging anticline looks the same as a cross-section of a non-plunging anticline. However, a cross-section parallel to the axis, such as the right-hand sides of the two block diagrams above, shows the beds dipping downward in the direction the fold is plunging.

In map view, a plunging syncline makes a U-shaped or V-shaped pattern that opens up in the direction of plunge.

Basins and Domes

A basin is a downward bulge in the rock layers, like a syncline without an axis. The beds all dip toward the center and the youngest rock is at the center.

A dome is an upward bulge in the rock layers, like an anticline without an axis. The beds all dip away from the center and the oldest rock is at the center.

Faults

A fault is a planar (relatively flat) surface within the earth, along which rocks have broken and slid. If a fault is not vertical, there are rocks on top of the fault and rocks beneath the fault. The rocks on top of a fault are called the hanging wall. The rocks beneath a fault are called the footwall.

Normal and Detachment Faults

In a normal fault, the hanging wall has slid down while the footwall has slid up along the fault.

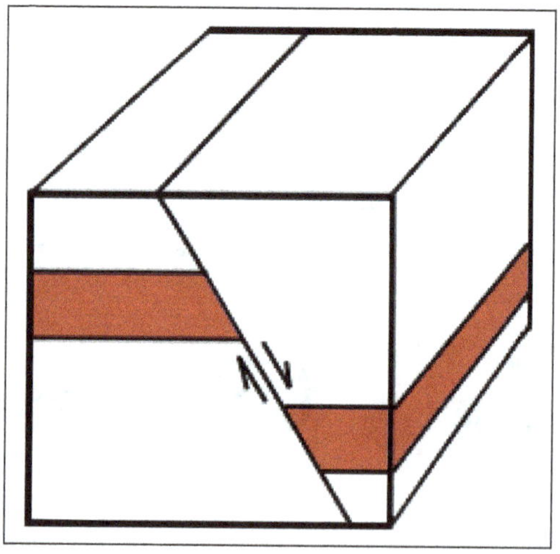

Normal Fault.

A detachment fault is a particular kind of normal fault with a low dip angle. It separates rocks that were deep in the crust (usually granite and gneiss), which were hot and underwent ductile deformation, from rocks of the upper crust (sedimentary or volcanic) that were cold and brittle. Detachment faults occur along the borders of metamorphic core complexes.

Normal and detachment faults form in sections of the crust that are undergoing tension-where the crust is being stretched apart. A divergent plate boundary is generally a zone of major normal faults, but normal faults also occur in other zones of crustal tension, such as in the Basin and Range landscape region.

Reverse and Thrust Faults

In a reverse or thrust fault, the hanging wall has moved up while the footwall has moved down along the fault. The difference between a reverse fault and a thrust fault is that a reverse fault has a steeper dip, more than 30°.

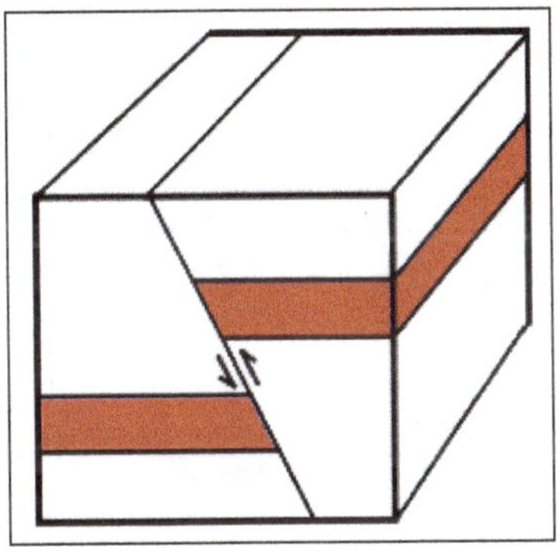

Reverse Fault.

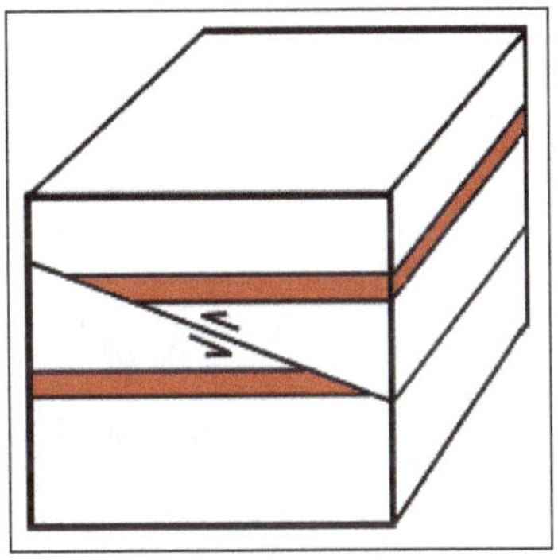

Thrust Fault.

Reverse and thrust faults form in sections of the crust that are undergoing compression. A convergent plate boundary is generally a zone of major reverse and thrust faults, but reverse and thrust faults also occur in other settings where the crust is being compressed.

Strike-slip Faults

Strike-slip faults are steep or vertical faults along which the rocks on either side have moved horizontally in opposite directions. A transform plate boundary is generally a strike-slip fault or zone of strike-slip faults, but strike-slip faults also occur in other settings.

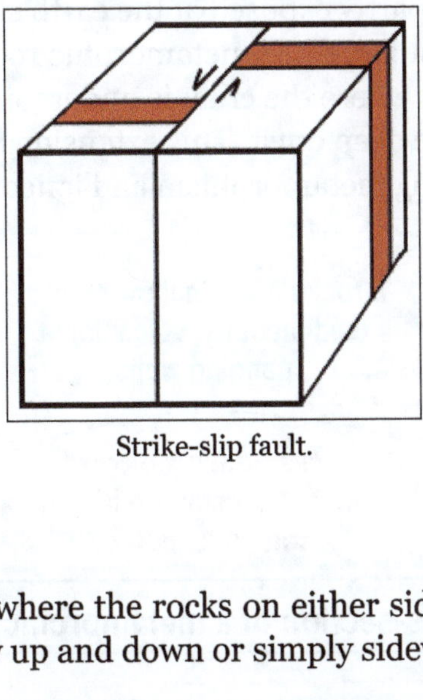

Strike-slip fault.

Oblique-slip Faults

An oblique-slip fault is one where the rocks on either side have moved up, down and sideways, rather than simply up and down or simply sideways.

Other Geologic Structures

Grabens and Horsts

A section of the crust that has dropped down along two parallel normal faults is called a graben. A block of crust that has been uplifted between two parallel normal faults is called a horst.

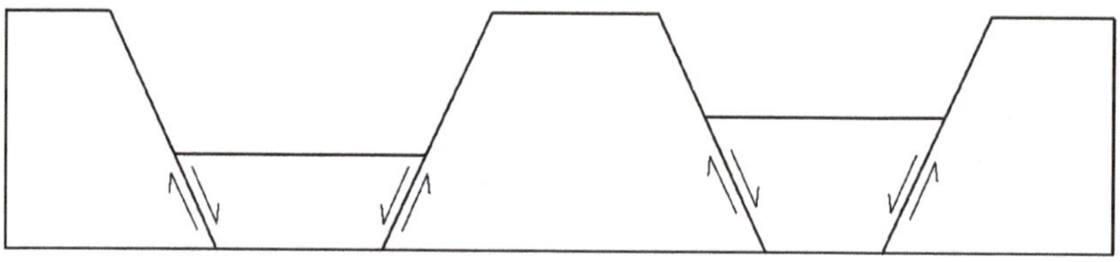

one complete horst and two grabens

Joints

Joints are the most commonly observed structure at the earth's surface. Near and at the earth's surface, rocks break and crack. Most often the rocks on either side of the crack do not slide or shift their position, so the cracks do not qualify as faults. Such cracks or fracture surfaces are called joints.

Joints are present in nearly every outcrop of bedrock at the earth's surface. Because it is common for joints to be relatively straight and flat they may be mistaken for sedimentary beds.

Metamorphic Core Complexes

A metamorphic core complex is a deep section of the crust that has uplifted along detachment faults and become partly exposed at the earth's surface. The rocks in a metamorphic core complex consist largely of metamorphic rock and granite. Metamorphic core complexes form in areas where the crust is undergoing extension, or being pulled apart, all the way down to the deep crust. This extension is occurring at the same time as the deep crust is undergoing metamorphism and igneous intrusion.

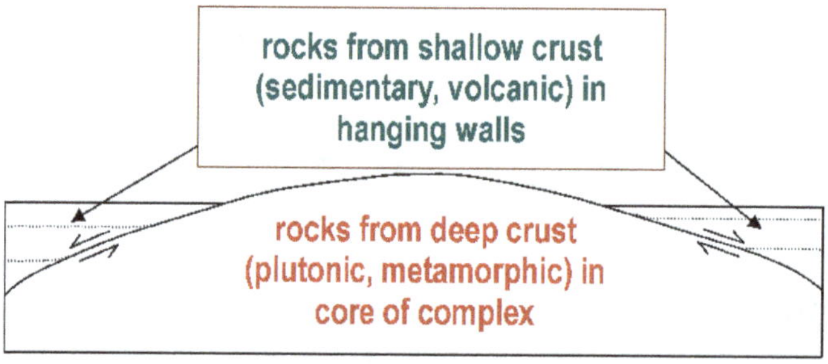

Simplified cross-section of a metamorphic core complex

Digital Elevation Models in Geomorphology

Geomorphology, the study of the Earth's physical land-surface features, such as landforms and landscapes and on-going creation and transformation of the Earth's surface, is one of the most important research disciplines in Earth science. The term geomorphology was first used to describe the morphology of the Earth's surface in the end of nineteenth century. Geomorphological studies have focused on the description and classification of landforms (geometric shape, topologic attributes, and internal structure), on the dynamical processes characterizing their evolution and existence and on their relationship to and association with other forms and processes.

Geomorphology dates back to sixth to fifth century BC, when Xenophanes of Colophon speculated that, as seashells are found on the top of mountains, the surface of the Earth must have risen and fallen, or Herodotus thought that the lower part of Egypt was a former marine bay referring to the year-by-year accumulation of river-borne silt in the Nile delta region. Geomorphology as an independent scientific discipline developed in the late nineteenth century. The first modern theory of landscape evolution was the 'geographical cycle' expounded by Davis. Another important theory, or rather a variation on Davis' scheme was offered by Penck who stated that according to the Davis model, uplift and planation take place alternately.

Geomorphology in the twentieth century experienced rapid evolution and growth, and many overlapping phases of development can be identified. Contemporary geomorphology combines and consists of many individual fields of science, that is, geology, hydrology, meteorology, cartography, geographic information system (GIS), engineering, biology, archaeology, etc. This complexity and the long tradition associated with its development helped geomorphologist to use many research techniques and measurement tools according to the research requirements. It is often said about the four main research directions, which are historical, dynamic, structural, and climatic, in geomorphology. However, as rightly observed by Migoń, each area has its own history, dynamic nature of the processes, and is present in specific geological structures and with the participation of specific climatic conditions.

- Morphography: It describes physical appearance of landforms and is qualitative approach to the form. It is linked with the direct observation of forms in situ which allows specifying the appearance of the form and its morphographic classification (plain, hill, valley, ridge, etc.). These terms do not indicate the way of creation of forms rather only determine their external expressions.
- Morphogenesis: It focuses on explaining the origin of the forms and determine the mechanisms of their contemporary development. Geomorphologists use different methods to determine the nature of the process in the past and the present form.

- Morphochronology: It aims to specify age of the forms and the age-relationships between adjacent landforms. Geomorphologists examine both absolute as well as relative age between the forms.
- Morphometry: It deals with establishing geometric features of the landforms on the basis of measurements.

The quantitative phase in geomorphology was developed in 1940–1970 and reflected a broader trend within many of the Earth sciences disciplines towards enhanced use of sophisticated technologies (often derived from military purposes) to measure, describe and analyse the Earth's surface features in number of categories. Horton's publications on stream networks and drainage basin processes are classically identified as the precursor to this quantitative movement. Strahler has limited a method for the quantitative analysis of the forms modelled by flowing water and gravitational movements over a longer period of time. The data were obtained mainly from measurements on detailed topographic maps and then applying statistical and morphometric analysis for various calculations and indicators of the fluvial relief within the drainage basin, karst, or glacial relief were made.

Further development of the geomorphometry resulted in working out of new theoretical and methodological basis. Lustig distinguished two approaches for quantitative analysis and Earth's surface characteristics which are (1) the characteristics of individual forms based largely on field surveys and (2) analysis of the surface area as a whole based on analysis of the map. Then Evans proposed the division of geomorphometry into two forms: General and specific. In general, geomorphometry applies to and describes the continuous land surface. It provides a basis for the quantitative comparison of qualitatively different landscapes and it can adapt to the methods of surface analysis used outside geomorphology. Whereas specific geomorphometry applies to discrete landforms, describes selected relief or landform types as well as their geometry and laws of formation and development. Although such divisions may seem somewhat artificial, since the geometry of individual forms consists of the geometry of the surface as a whole, but they are useful to define more closely the principles of various types of morphometric analysis.

At the end of the 1980s and the beginning of the 1990s, with the personal computer revolution, algorithms have been implemented in many raster-based GIS packages (ArcInfo, MicroStation, MicroDEM, etc.) and it was possible to process digital elevation models (DEMs) over fairly large areas. Then began the new era of geomorphometry with new opportunities to visualize and compute land-surface parameters. Nowadays, geomorphometry is an important component of terrain analysis and surface modelling including measurements of morphometry of continental ice surfaces, characterizing glacial troughs, mapping sea-floor terrain types, guiding missiles, assessing soil erosion, analysing wildfire propagation, and mapping ecoregions; it is a broad field that is important not only in various aspects of Earth sciences but also in engineering, biology, and medicine.

Digital Elevation Models (DEMs)

A numerical description of the ground surface is helpful in addressing many geomorphological problems. In the definition of the subject of this chapter appears the word 'model', that is, a representation, generally in miniature, to show the construction or appearance of something. Meyer neatly expressed it—reality scaled down and converted to a form which we can comprehend. In this case it is used to represent the original situation—approximation of topography. The term and concept of 'terrain model' was first described by Miller and Laflamme but it did not come into general use until the 1960s and even later because of the technological limitations (computers, processors, etc). They defined digital terrain model (DTM) as just a statistical representation of continuous Earth's surface which consists of many points of known co-ordinates x, y, z in the arbitrary co-ordinate system. In the following years, a number of terms related to the presentation of the Earth's surface by numerical methods, that is, Burrough described DEM as a regular gridded matrix representation of the continuous variation of relief over space. Among many terms, the most common and accepted in geomorphometric and GIS terminologies is digital elevation model (DEM) or digital terrain model (DTM). Some authors make distinctions between them, that is, DEM is 'an ordered array of numbers that represent the spatial distribution of elevations above some arbitrary datum in a landscape'. DTM is 'an ordered array of numbers that represent the spatial distribution of terrain attributes'; therefore DEM is a subset of DTM. Moore et al., Weibel and Heller, and Li et al. defined DTM as a digital (numerical) representation of the terrain. Herein both terms will be treated as synonyms but, in fact, the concept of DEM is broader and more universal. In summary, we can say that digital elevation model is the set of digital data describing elevation values of Earth's ground surface (or any other surface) which contains additional information about the character of this surface (i.e. structural lines, break lines, water bodies, etc.) and interpolation algorithm, which is the best for approximation (modelling) of the real topography. A DEM is a complete representation of a land surface which means that heights are available at each point in the area of interest.

Types of Digital Elevation Models

Due to geometry and data organization there are several basic types of models: Raster (GRID), vector triangulated irregular network (TIN), point and hybrid.

Raster model (GRID) is the most widely used digital height data structure during the last years because of its simplicity. The simplicity consists, in fact, that elevations are stored using a regular square grid that is consistent in each part of the study area. All square grids form regular matrix of heights for which plain co-ordinates (x, y) can be easily calculated due to the regular spacing of the grid points. This kind of DEM has many advantages, such as simple elevation matrices that record topological relations between data points implicitly and ease of computer implementation. Moreover, DEM is considerably easier to design land-surface parameters and objects using grids because simpler algorithms can be used; grids have a uniform spatial structure and almost all properties of gridded DEMs are defined by a single characteristic which is cell size; a grid model is more suited to the computer models used in image processing and for printing. For this reason, some DEM software packages accept only grid data.

Example of the GRID model (A), and TIN model (B).

Raster DEMs also have disadvantages, such as grids present under-sampled topography in areas where the topography is complex, and they over-sample smooth topography; re-projection of a grid is slow and sometimes leads to a loss of accuracy (because the initial grid loses its regular structure in a new projection and so it has to be re-calculated); the different distances between grid centres in cardinal and diagonal directions have a negative impact on the precision of many hydrological modelling; the computed upslope flow paths will tend to have zigzag pattern across the landscape and increase the difficulty of calculating specific catchment areas accurately; square grids cannot handle abrupt changes in elevation easily and they will often skip important details of the land surface in flat areas.

The second most popular and widely used is a vector model named triangulated irregular network (TIN). TIN is proposed by Hormann which devised a TIN idea, linking selected points on divides, drainage lines and breaks in slope to interrelate height, slope gradient and aspect. TIN is based on triangular elements (facets that vary in shape and size) with vertices at the sample height points. These facets consist of planes joining the three adjacent points in the network and are usually constructed using Delaunay triangulation. TIN can be interpolated directly from surveyed points or discrete features that are extracted manually from maps or by computer from a grid or contour DEM. The triangle may be regarded as the most basic and universal unit in all geometrical patterns (because of their great flexibility in terms of shape and size), since a regular grid of square or rectangular cells or any polygon with any shape can be decomposed into a series of triangles.

The advantages of such a model are: relevance to gravitational movements, and especially to hydrological applications; TINs are widely used in perspective representations of surfaces, especially in dynamic fly-through displays where the foreground is represented in full detail but the background can be simplified as larger triangles; for areas with high relief or rougher surfaces, irregular DEMs can use smaller spacing between points, and larger spacing where relief is lower or where the surface is smoother and in this way they can more accurately describe geological faults and other sharp elevation changes using the same number of points as grids; the TIN structure gives best reflecting processes of erosion and deposition mimics paths of steepest gradient. Of course we can use the algorithms to convert the TIN into grid and vice versa, but usually at the loss on the quality and accuracy of the model.

Data Sources

DEM source means the data acquired by different techniques and from different sources. In general, such data can be acquired from topo maps and from field surveys. Manual or semi-automated digitizing of contour lines and every single elevation point on topographic maps on the computer screen or digitizer can be done relatively cheaper without the need for resurveying. Derived DEMs represent the underlying terrain without surface vegetation and buildings. Despite many critical voices regarding the use of this method, it is a quick and effective method to get quite good DEM and till the end of 1990s this was the most common source of elevation data for DEMs.

The second group of data sources are surveys, such as ground survey techniques, digital photogrammetry, and remote sensing. Field surveys are carried out using total stations, theodolite, levelling instruments, global positioning system (GPS), and differential global positioning system (DGPS). These kinds of surveys have high accuracy (sometimes even less than 1 cm), flexibility (the measurement density can be varied, depending on the terrain), and very little processing is required after the measurements have been taken, but especially suited for measuring small areas. Digital photogrammetry relies on the stereoscopic interpretation of aerial photographs or satellite imagery using manual or automatic stereoplotters. Remote sensing surveys consist of Airborne laser scanning (ALS) or satellite platforms and include laser-ranging altimetry (LiDAR—light detection and ranging), synthetic aperture radar interferometry (InSAR or IfSAR). The final results will often include tree-top canopies and buildings. This gives higher elevation values, rough surfaces, and high slope values. The advantages of the LiDAR method are production time that is typically shorter than that for photogrammetrically generated DEMs and great spatial and vertical accuracy (<0.5 m). The main disadvantage of LiDAR data is point cloud which produces a very dense and detailed land-surface model that could be difficult to handle during the production process and sometimes the accuracy of the readings vary according to the characteristics of the terrain (i.e. very steep slopes). However, these days LiDAR is definitely the best method of the DEM production. Several countries have already produced national LiDAR DEM/DSM (e.g. Belgium, the Netherlands at resolutions of 2–5 m, Poland at 1 m).

Free Global DEMs

Currently, many elevation data with better accuracy and parameters are freely available in the world including LiDAR data. Individual countries or institutions offer different kinds of models.

GTOPO30 (Global 30 arc-second Elevation): This is a global DEM with a horizontal grid spacing of 30 arc seconds (approx. 1 km) resulting in a product having dimension of 21,600 rows and 43,200 columns. GTOPO30 was derived from eight several raster and vector sources of topographic information which are digital terrain elevation data (DTED), Digital Chart of the World, USGS 1-degree DEM's, army map service (AMS) 1:1,000,000, International Map of the World 1:1,000,000, Peru Map 1:1,000,000, New Zealand DEM, and Antarctic Digital Database. The horizontal accuracy is ±150 m linear error at 90% confidence and vertical accuracy is 10–300 m at the 90% confidence level.

GLOBE DEM (Global Land 1-km base elevation DEM): It is one of the first global DEM, which was initially spearheaded in 1990. This data set covers 180°W to 180°E longitude and 90°N to 90°S latitude. The horizontal resolution is 0.5 arc-minute in latitude and longitude, resulting in dimensions of 21,600 rows and 43,200 columns. GLOBE version 1.0 has 11 broad sources of information: 6 gridded DEMs and 5 cartographic sources, which were adapted for use in GLOBE. These were digital terrain elevation data (DTED), Digital Chart of the World, Australian DEM, Antarctic Digital Database, Brazil Maps 1:1,000,000, DEM for Greenland, army map service (AMS) Maps 1:1,000,000, DEM for Japan, DEM for Italy, DEM for New Zealand, and Peru Map 1:1,000,000. Vertical accuracy expressed as ±30 m linear error at 90% confidence can also be described as a root mean square error (RMSE) of 18 m. Linear error distribution is 97 m for RMSE and it can be expressed as ±160 m linear error at 90% confidence.

Shuttle radar topography mission v3 (SRTM): It was flown to aboard the space shuttle Endeavour February 11–22, 2000. The National Aeronautics and Space Administration (NASA) and the National geospatial-intelligence agency (NGA) participated in an international project to acquire radar data that was used to create near-global set of land elevations. SRTM successfully collected radar data over 80% of the Earth's land surface between 60°N and 56°S latitude with data points posted every 1 arc-second (approximately 30 m). Absolute height error of SRTM data sets is 5–10 m and absolute geolocation error is 7–13 m. The level of processing and the resolution of the data vary by SRTM data set:

- SRTM Non-Void Filled: This version was edited or finalized by the NGA to delineate and flatten water bodies, better define coastlines, remove spikes and wells, and fill small voids. Data were sampled at 1 arc-seconds (USA) and 3 arc-seconds (rest of world).
- SRTM Void Filled: This elevation data are the result of additional processing to

address areas of missing data or voids in the SRTM Non-Void Filled collection. The resolution for SRTM is 1 arc-seconds (USA) and 3 arc-seconds (rest of world).

- SRTM 1 Arc-Second Global: This elevation data offer worldwide coverage of void filled data at 1 arc-second (30 m) resolution and provides open distribution of this high-resolution global data set. Some tiles may still contain voids. Please note that tiles above 50°N and below 50°S latitude are sampled at a resolution of 2 by 1 arc-second.

ETOPO1 (Global Relief Model at 1 arc-minute resolution): It was made in 2008 by the National Geophysical Data Centre (NGDC), an office of the National Oceanic and Atmospheric Administration (NOAA). This model was developed as an improvement to the ETOPO2v2 Global Relief Model. ETOPO1 is available in two versions: 'Ice Surface' (top of Antarctic and Greenland ice sheets) and 'Bedrock' (base of the ice sheets). These versions of ETOPO1 were generated from diverse (regional and global) digital data sets, which were shifted to common horizontal and vertical datum. Next steps were evaluated and edited as needed. Shoreline, bathymetric, topographic, integrated bathymetric–topographic, and bedrock digital data sets (13 sources) were obtained from several U.S. government agencies, international agencies, and academic institutions.

ASTGTM (ASTER global digital elevation model v002 and AST14DEM—ASTER digital elevation model v003): These DEMs were developed jointly by the US NASA and Japan's Ministry of Economy, Trade and Industry (METI). ASTER is capable of collecting in-track stereo using nadir and aft-looking near infrared cameras. Since 2001, these stereo pairs have been used to produce single-scene (60 × 60 km) DEMs having vertical RMSE accuracies generally between 10 and 25 m. The ASTER GDEM covers land surfaces between 83°N and 83°S and is comprised of 22,702 tiles. This model is distributed as georeferenced tagged image file format (GeoTIFF) files. The data have resolution at 1 arc-second (approximately 30 m) grid and referenced to the 1984 World Geodetic System (WGS84)/1996 Earth Gravitational Model (EGM96) geoid. Although the ASTER GDEM v. 002 is better model than ASTER GDEM v. 001, users have to know that the data still may contains anomalies and artefacts. One should know that these mistakes can introduce large elevation errors on local scales.

GMTED 2010 (Global Multi-resolution Terrain Elevation Data): This model was made by collaborating USGS and the NGA, which replaced GTOPO30 model. The new elevation data set has been generated at three separate horizontal resolutions of 30 (about 1 km), 15 (about 500 m), and 7.5 arc-seconds (about 250 m). The global aggregated vertical accuracy of GMTED2010 can be summarized in root mean square error (RMSE). At 30 arc-seconds, the RMSE range is 25–42 m; at 15 arc-seconds, range is 29–32 m, and at 7.5 arc-seconds, range is in between 26 and 30 m. This new product suite provides global coverage of all land areas from latitude 84°N to 56°S for most products, and coverage from 84°N to 90°S for several products. An additional advantage of the new multi-resolution model over GTOPO30 is that at each resolution seven new raster elevation data sets are available. The new models have been produced using the various aggregation methods: minimum elevation, maximum elevation, mean elevation, median elevation, standard deviation of elevation, systematic subsample, and breakline emphasis. GMTED2010 is based on data derived from different raster-based elevation sources, such as SRTM DTED, non-SRTM DTED, Canadian digital elevation data (CDED), Satellite Pour l'Observation de la Terre (SPOT 5), Reference3D, National elevation dataset (NED), GEODATA, an Antarctica satellite radar and laser altimeter DEM, and Greenland satellite radar altimeter DEM.

AW3D30 (ALOS Global Digital Surface Model): The global digital surface model (DSM) at 1 arc-second (approximately 30 m) resolution that was released by Japan aerospace exploration agency (JAXA). This model has been compiled with images acquired by the advanced land observing satellite (ALOS). The elevation data are published based on the DSM data set (5-m mesh version) of the 'World 3D Topographic Data'. The source data were a huge amount of stereo-pairs images derived from satellite mission in the years 2006–2011. Next, they were processed semi-automatically to provide digital surface model (DSM). The height accuracy of the data set is approximately <5 m from the evaluation with ground control points (GCPs) or reference DSMs derived from the LiDAR.

In addition to the above global DEMs, there are many elevation data with much better accuracy, but for smaller areas, for example, for Europe EU-DEM with grid size 25 m, for Spain MDT05/MDT05-LIDAR (5 m), for Netherlands AHN3 DTM (0.5 m), for all Slovenia (LiDAR data), or LiDAR data for Haiti (1 m).

Morphometric Landform Properties

The main object of the studies in geomorphology is the relief (of the Earth's surface). The term relief of the Earth's surface was used to describe the vertical dimension or amplitude of topography or as the elevation difference over a pre-determined area or collective elevations and their inequalities of a land surface. One can also say that the relief is the complexity of the Earth's surface shapes, and these shapes form the landforms. If we would present a continuous Earth's surface as a matrix of discrete points with a defined distance interval from one another, the shape dimensions would express in the changes of height, distance, and direction between these points. These three vectors and the relationships between them are the basis for most morphometric landform properties.

Morphometric Parameters

Elementary data used in geomorphometry are DEM. Strength of DEM is a suggestive (plastic and three-dimensional) visual message and the ability of quantifying the topography. Quantification of the Earth's surface is expressed by topographic attributes which can be determined from the derivatives of the topographic surface. Generally, one can say, these derivatives measure rate at which elevation changes in response to changes in location. Deng et al. noted that local terrain shape (as the continuous variation of elevation values over the terrain surface from point to point) has an enormous impact on local terrain attributes, but this role is influenced by data and computational factors.

Theoretical assumptions concerning the morphometric properties of the surface come from the 1950s, and even earlier. Stone and Dugundji and Hobson stated that measures of landforms can be considered as a kind of the roughness of the surface. In general, roughness refers to the irregularity of a topographic surface and cannot be completely defined by any single measure but must be represented by a roughness vector or set of parameters. With roughness, concept of wavelength and amplitude ideas was related. The significant wavelengths of topography are termed grain or texture, while amplitudes associated with these wavelengths correspond to the concept of relief. Texture and grain are terms that have been used to indicate in some way the scale of horizontal variations in the topography. Texture is used to refer the shortest significant wavelength in the topography and grain is used for the longest significant wavelength. Texture is related to the smallest landform elements, while the grain is related to the size of area over which one measures other parameters. Wood and Snell defined grain as the size of area over which the other factors are to be measured. It is dependent on the spacing of major ridges and valleys and thus indicates texture of topography. Texture refers to the shortest significant topographic wavelength.

The systems proposed by Evans and Krcho for field variables are local-based, relating to a vanishingly small area around each point. They divided topographic properties into primary geomorphometric parameters (height, slope, aspect, profile, and planar curvature) and statistical measures derived in square-matrices out of the gridded DEMs (such as relief, standard deviation, and skewness of heights).

Mark stated, that probably the most important single class of processes which has shaped the Earth's surface could be divided into geometrical properties that involve the relationships among dimensional properties, such as elevation, lengths, areas, volumes, and topological properties which relate numbers of objects in the drainage network (e.g. the bifurcation ratio). Then Moore and Thornes developed LEAP-land erosion analysis programs to examine the spatial distribution of slope length, slope steepness, and the plan curvature for assessing topographic erosion potential.

Pike noted that terrain can be abstracted using geometric signature—a set of measurements that describes topographic form sufficiently well to distinguish geomorphically disparate landscapes. Then he developed the concept of a geometric signature, a multi-variate description of topography using a suite of measures, and later expanded the concept with a listing of 49 variables that could be grouped into 22 attributes. He considered roughness and height, the two most important attributes, with two measures of texture at seventh and eleventh position. Fifteen different variables contribute to roughness. Pike et al. also referred to grain concept. Relief at topographic grain is an estimate of local relief optimized by varying unit-cell size. In homogeneous terrain, local relief within nested circles increases with circle size and then levels off at a diameter termed grain, a measure of characteristic local ridgeline-to-channel spacing. Topographic grain is the characteristic horizontal spacing of major ridges and valleys and is an important descriptor of meso-scale topographic texture. The grain concept arose from the need for a variable and non-arbitrary unit-cell size and also enables to calculate local relief parameter.

They divided terrain attributes into primary and secondary (compound) attributes. Primary attributes, hydrologically related, are directly calculated from elevation data and include variables, such as altitude, upslope height, aspect, slope, upslope slope, dispersal slope, catchment slope, upslope area, dispersal area, catchment area, specific catchment area, flow path length, upslope length, dispersal length, catchment length, profile curvature, and plan curvature. It is interesting that Bork and Rohdenburg have developed a digital relief model for estimating and displaying the distribution of these morphographic parameters.

Schmidt and Dikau subdivided primary geomorphometric parameters into three types: simple, complex, and combined. Simple primary geomorphometric parameters (usually derived through a filter operation within a 3 × 3 moving window) are height, slope, aspect, profile curvature, contour curvature, drainage direction, and real area of pixel. Complex geomorphometric parameters are derived through the analysis of the whole matrix of a DEM. They contain structural information about the surrounding morphometry and consist of contributing area, mean slope of contributing area, average, and variance of primary parameters in contributing area, length of flowpath to outlet, average and variance of primary parameters in flowpath to outlet, length of flowpath to stream, average and variance of primary parameters in flowpath to stream, x- and y-co-ordinates of corresponding stream point, height of corresponding stream point, height distance to corresponding stream point, length of minimum flowpath to watershed, relative slope position after minimum, slope length, relative slope position after maximum, and minimum and maximum slope length.

Interesting approach to the morphometric parameters was presented by Shary et al., who stated that morphometric variables describe not the land surface itself, but rather the system, land surface + vector field, where vector fields of common interest are gravitational field and solar irradiation. He divided morphometric variables and concepts into field-specific (may refer to this system description) and field-invariant (invariant with respect to any vector field, that is, describing the land surface geometrical form). On the other hand, morphometric variables may divide into local, regional, or global (when height data of all the Earth is needed for their determination).

Basso divided topographic attributes into: (i) local (calculated from a small neighbouring area surrounding the DEM cell, usually 3 × 3), (ii) regional (calculated using considerably larger geometric area than the local attributes, less sensitive to the DEM resolution), (iii) catchment oriented (related to the whole catchment area, and are the measurement of certain catchment characteristics), and (iv) process oriented (describe or characterize the spatial variability of a simple representation of specific processes that occur on the landscape).

Goodwin and Tarboton described five categories of the morphometric parameters of drainage basin which are size properties (provide measures of scale that can be used to compare the magnitudes of two or more drainage basins), surface properties (quantities depicted by fields comprising a value at each point within a domain, drainage basin), shape properties (i.e. length, width, perimeter, and more complex function), relief properties (total basin relief, relief ratio, and hypsometric curve), and texture properties (amount of landscape dissection by a channel network).

Olaya presented division of the morphometric parameters into local and regional. Local parameters consist of geometric (slope, aspect, curvatures, visibility, visual exposure, and visibility index) and statistical (i.e. average value, standard deviation, skewness coefficient, kurtosis coefficient, range of values, etc.). Regional parameters are connected with hydrological properties (catchment area, height, slope, proximity, etc.).

The latest approach to the classification system of the fundamental geomorphometric variables is presented by Evans and Minár. They proposed field and object variables. Field variables include specific to gravity field (local point-based, local area-based, and regional), specific to other fields and field-invariant variables (local point- and regional-based). Object variables differ between areal, linear, and point features.

Thus the optimal number of variables depends on spatial scale, resolution of the source data (DEM), and the requirements of the research problem; and many measures describe the same attribute of surface form and thus are redundant; for this reason new geomorphometric parameters are very rare.

Morphometric Indices

Morphometric parameters, showed that there are many different classifications of these parameters. We want to look at some popular morphometric indices called secondary or composite attributes, combined or compound geomorphometric parameters, statistical parameters or process oriented. These indices are combinations of the primary morphometric attributes and describe or characterize the spatial variability of specific processes occurring on the landscape. Sometimes these morphometric indices can be derived empirically but it is preferable to develop them through the application and simplification of the underlying physics of the processes. With the index approach we simplify some physical sophistication to allow improved estimates of spatial patterns in the landscape.

Morphometric Tools available in GIS Packages

The current availability of high speed computing platforms and high-resolution (less than 10 m spatial resolution) DEMs now provides the opportunity to perform quantitative analyses and calculating morphometric indices on new level. Many GIS packages offer tools to work with DEMs. On one hand we have the comWmercial software (ArcGIS, MapInfo, Surfer, Global Mapper, Terra Solid), and on other one, great free programs (SAGA, QGIS, MicroDEM), which offer a lot of interesting tools.

Landform Classifications

Geomorphology studies the relief. If we want to understand the relief of the Earth's surface (which is highly complex) we need to simplify and subdivide it into landforms. We have to focus on describing landforms, their spatial arrangement and the processes which led to their formation. Of course, many landforms can be delineated manually using photo-interpretation to assess their form, size, scale, adjacency, surface roughness, hydrological and contextual position but there is always problem with the boundary of the landform. And herein DEMs are helpful. Availability of global DEMs and high accuracy national LiDAR data made new possibilities of analysing these data to extract and classify geomorphic entities. The landform elements can be extracted automatically from DEMs by using land-surface parameters, such as slope, curvatures, catchment area, distance to streams, peaks and depression depth, etc. The goal of automated extraction of landforms and landform elements using semi-automated or fully-automated algorithms is to find their geometric signature, which Pike defined as a set of measures that describe topographic form well enough to distinguish among geomorphically disparate landscapes.

There are many landform classification systems, and herein we take this issue very briefly and in general. The pioneer works in quantitative systems for landform classification were conducted by Hammond, Wood and Snell and Anstey by using topographic contour maps. These studies were aimed primarily towards the systematization and logical interpretation of terrain data to assist in the determination of design criteria for materiel, and secondarily towards the development of a universal system for the quantification of landform data.

In the mid-1960s, Hammond devised the three-level system of regional landform classification, based purely on geomorphometric parameters calculated in the chosen window size. For each window position the following parameters were calculated: (1) percentage of area where the ground is flat or gentle (less than 8% slope); (2) local relief (maximum minus minimum elevation); and (3) profile type (relative proportion of flat or gently sloping terrain that occurs in lowlands or uplands).

Peucker and Douglas showed a few methods designed to detect pits, peaks, passes, ridges, ravines, and breaks, given an array of sampled, quantized terrain heights. Described methods used local analysis which means the results at each point do not depend on the results already achieved at other points.

As the hillslopes constitute a basic element of all landscapes and a fundamental component of geomorphologic systems, many subjective classifications of slope profiles that were intended to create conceptual classifications of hillslopes have been proposed. Ruhe divided hillslopes into summits, shoulders, backslopes, footslopes, and toeslopes. Dalrymple et al. and Conacher and Dalrymple proposed nine unit classification of hillslopes: (i) Interfluve, (ii) Seepage slope, (iii) Convex creep slope, (iv) Fall face, (v) Transportational midslope, (vi) Colluvial footslope, (vii) Alluvial toeslope, (viii) Channel wall, and (ix) Channel bed (after).

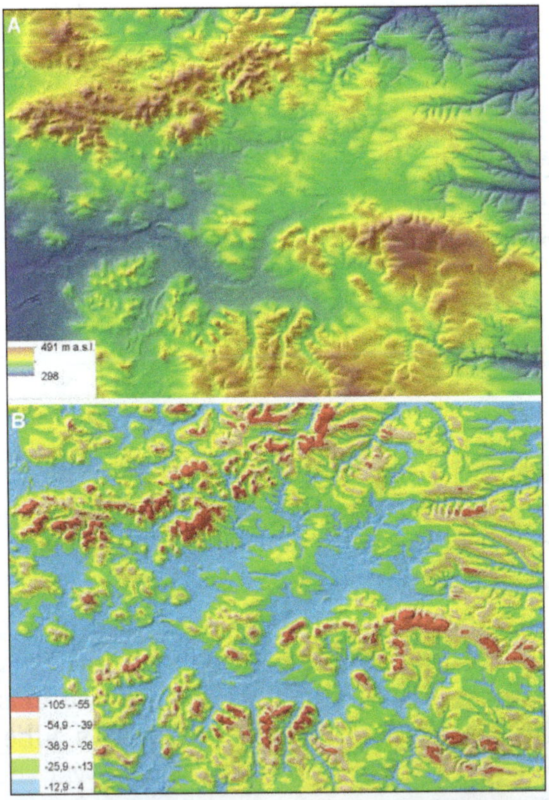

Elevation map (A), and simplified Hammond's landform classification (B).

Next, there were several approaches to standardize hillslope units using qualitative terms. Most commonly, a hillslope was described by a series of basic units describing changes in slope, curvature, and processes along the hillslope profile.

As Young noted the curvature can be classified into convex, concave, and rectilinear surfaces, Dikau defined basic form elements of the landscape as the combination of three slope profile curvature characteristics and three plan curvature characteristics which lead to nine possible hillslope units. They deliver a disjunctive description of the hillslope surface into units of homogeneous curvature characteristics. Then Dikau proposed digital geomorphological relief model (DGRM) which will generate form facets and elements (as basic relief units) for geomorphological mapping and simulations for the derivation of more complex relief units. Dikau et al. developed and applied an automated method for classifying macro landform types from DEM that was based on analysis of variation in topographic measures within areas defined by moving windows.

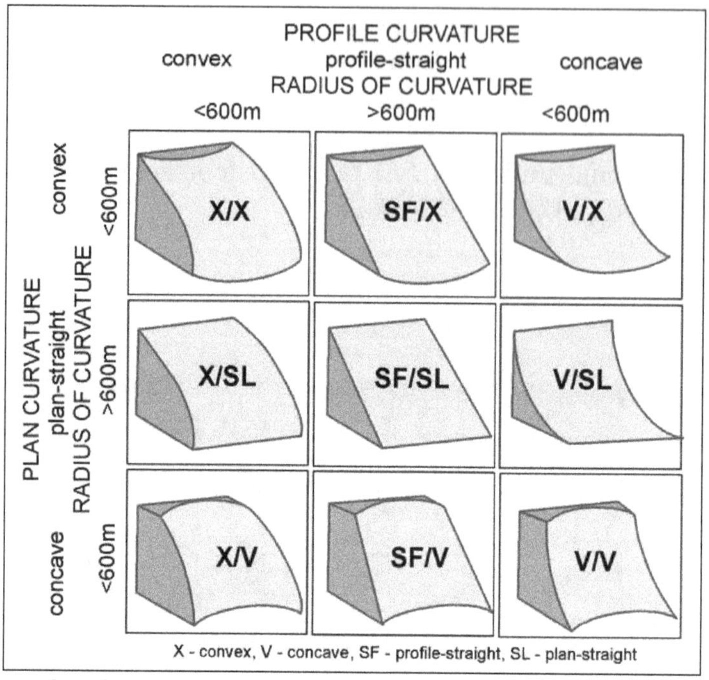

Classification of form elements by plan and profile curvature.

The original classification explicitly assumed that surface form, as described by curvature, could be directly related to surface processes and to relative landform position (divergent/convergent shoulder, backslope footslope and level). Thus, strong profile convexity was assumed to be indicative of upper, water-shedding slope positions; whereas strong profile concavity was associated with lower, water-receiving landform positions, and planar surfaces were associated with backslopes or flat areas. Of course, this pattern is not always adhered to and there are many instances where convex-concave patterns repeat over short distances along a longer hill slope.

Speight described 10 morphological types of topographic landform positions: (i) Crest, (ii) Depression (open, closed), (iii) Flat, (iv) Slope, (v) Simple slope, (vi) Upper slope, (vii) Mid-slope, (viii) Lower slope, (ix) Hillock, and (x) Ridge (after).

Very simple and interesting method for classifying relief on the base DEM is topographic position index (TPI). TPI is the difference between the elevation at a cell and the average elevation in a neighbourhood surrounding that cell. Positive values means that the cell is higher than its neighbours (indicate ridges, hills, etc.) while negative values means the cell is lower (indicate canyons, valleys, etc). TPI is a simplification of the landscape position index (LPI) described by Fels and Zobel. TPI values provide a simple and powerful means to classify the landscape into morphological classes. TPI is naturally very scale-dependent, it means neighbourhood size (and shape) and DEM resolutions are critical to the final analysis, so the work with this index should be based on experiments with different threshold values.

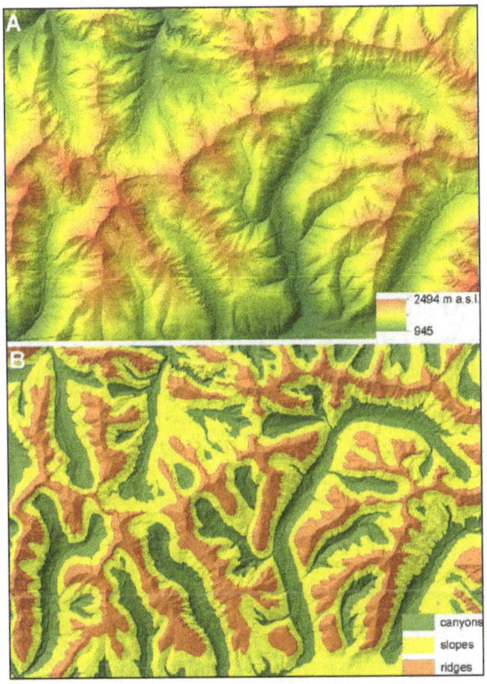

Elevation map (A), and TPI 3-category slope classification (B).

Drăguţ and Blaschke presented an automated classification system of landform elements based on object-oriented image analysis. Firstly, one need to derive elevation, profile curvature, plan curvature, and slope gradient from DEM. Next, relatively homogenous objects are determined at several levels through segmentation of the image. These object primitives are classified as landform elements using a relative classification model, built both on the surface shape and on the altitudinal position of objects. The classification has nine classes which are peaks and toe slopes, steep and flat/gentle slopes, shoulders and negative contacts, head slopes, and side and nose slopes. Classes are defined using flexible fuzzy membership functions.

Iwahashi and Pike developed an iterative procedure that classifies topography automatically into terrain types, grid cell by grid cell, on the basis of three morphometric variables, such as slope gradient, local convexity, and surface texture. They applied an unsupervised nested-means algorithm for 16 topographic types of the world. They noted the procedure is unsupervised and reflects frequency distributions of the input variables but not defined criteria. It causes that resulting classes are undefined and have to be calibrated empirically by subsequent analysis.

Within the context of defining landform units that maximize internal homogeneity and external differences, Minár and Evans presented the concept of elementary forms (segments and units) defined by constant values of fundamental morphometric properties and limited by discontinuities of the properties. The basic system of form-defining properties represents altitude and its derivatives, constant values of which provide elementary forms with various types of homogeneity.

Drăguţ et al. presented an algorithm to derive elementary forms from DEMs. Elementary forms were defined by constant values of fundamental morphometric properties and limited by discontinuities of these properties. A multi-resolution segmentation technique was customized to partition the layers of altitude derivatives into homogeneous divisions through a self-scalable procedure, which reveals the pattern encoded within the data. Layers were segmented successively, following the order of elevation derivatives (i.e. gradient, aspect, profile curvature, and plan curvature).

Jasiewicz and Stepinski introduced a novel method for unsupervised classification and mapping of landforms from a DEM. This method involves the pattern recognition, not the differential geometry. The foundation of this idea is the concept of geomorphon (geomorphologic phonotypes), which is a simple ternary pattern that serves as an archetype of a particular terrain morphology. A finite number of 498 geomorphons constitute a comprehensive and exhaustive set of all possible morphological terrain types including standard elements of landscape as well as unfamiliar forms rarely found in natural terrestrial surfaces.

And one should remember that horizontal and vertical resolution of the elevation data used to present a terrain surface has a significant influence on the level of detail and the accuracy of portrayal of surface features and on the values of land-surface parameters that are computed from a DEM. One should first test out predictive efficiency for various DEM resolutions and neighbourhood sizes, and then objectively derive the most suitable resolution and search size.

Branches of Geomorphology

2

- **Fluvial Geomorphology**
- **Biogeomorphology**
- **Climatic Geomorphology**
- **Hydrogeomorphology**
- **Applied Geomorphology**

Geomorphology can be categorized into a varied number of branches. Some of them are fluvial geomorphology, biogeomorphology, climatic geomorphology, hydrogeomorphology, applied geomorphology, etc. The topics elaborated in this chapter will help in gaining a better perspective of the different branches of geomorphology.

Fluvial Geomorphology

Fluvial geomorphology is the study of the interactions between the physical shapes of rivers, their water and sediment transport processes, and the landforms they create.

It studies the ways that rivers move and change over time, focusing especially on how the flow of water interacts with the movement of sediment – dirt, sand, gravel, boulders – and debris, such as downed trees and branches. It also considers how the movement of water, sediment and debris interacts with the fixed, immobile features of the landscape, from bedrock canyons to human-built infrastructure like dams, bridges, and reinforced stream banks.

The starting points for understanding how rivers move and interact with their landscapes are:

- Rivers include sediment and debris as well as water; and as they flow, they apply force on, and release material to, the landscapes around them.
- As the water in a river travels, its pushes on the rocks, sands and silts in its bed and bank. Often it dislodges some of these sediments, and carries them into the channel and downstream. When the water is fast and voluminous, a river can erode and carry a lot of sediment, and bigger pieces.
- When a river slows or becomes shallow, sediments settle out.

The higher the stream power the larger the material size that a stream is able to move. Stream power is determined by the amount of water flowing through a stream (discharge) and the velocity at which that water is flowing. Generally, steep channels contain powerful streams that erode large materials. Low-gradient channelshave low-powered streams that deposit fine materials.

Biogeomorphology

Slope stabilization by Chilean rhubarb on the coasts of Chacao Channel. Vegetation have mostly a protective effect on slopes.

Biogeomorphology and ecogeomorphology are the study of interactions between organisms and the development of landforms, and are thus fields of study within geomorphology and ichnology. Organisms affect geomorphic processes in a variety of ways. For example, trees can reduce landslide potential where their roots penetrate to underlying rock, plants and their litter inhibit soil erosion, biochemicals produced by plants accelerate the chemical weathering of bedrock and regolith, and marine animals cause the bioerosion of coral. The study of the interactions between marine biota and coastal landform processes is called coastal biogeomorphology.

Phytogeomorphology is an aspect of biogeomorphology that deals with the narrower subject of how terrain affects plant growth. In recent years a large number of articles have appeared in the literature dealing with how terrain attributes affect crop growth and yield in farm fields, and while they don't use the term phytogeomorphology the dependencies are the same. Precision agriculture models where crop variability is at least partially defined by terrain attributes can be considered as phytogeomorphological precision agriculture.

Biogeomorphology is a multidisciplinary focus of geomorphology that takes research approaches from both geomorphology and ecology. It is a sub discipline of geomorphology. Biogeomorphology can be synthesized into two distinct approaches:

- The influences that geomorphology plays on the biodiversity and distribution of flora and fauna.
- The influences that biotic factors have on the way landforms are developed.

There has been much work on these approaches such as; the effect that parent material has on the distribution of plants, the increase of precipitation due to an influx of transpiration, the stability of a hillslope due to the abundance of vegetation or, the increase of sedimentation due to a beaver dam. Biogeomorphology shows the axiomatic relationship between certain land forming processes and biotic factors. That is, certain geomorphic processes shape the biota and biotic factors can shape land forming processes.

Darwin's work on earthworms, the caption of which reads "A tower-like casting, probably ejected by a species of Perichæta.

The earliest work related to biogeomorphology was Charles Darwin's 1881 book titled The Formation of Vegetable Mould through the Action of Worms. Although the field of biogeomorphology had not yet been named, Darwin's work represents the earliest examination of a faunal organism influencing landscape process and form. Charles Darwin begins his work on worms with an examination of behavior and physiology, which then moves towards topics related to geomorphology, pedogenesis, and bioturbation. Observations and measurements of soil moved by earthworms, and emphasis on the role of earthworms in formation of humus, fertility of soils, and mixing of soils were all described in the book, which began to change the perspective on earthworms from pest to critical agent of pedogenesis. Despite the popularity of Darwin's final work, the scientific community was slow to recognize the significance of examining the role of organisms in influencing landscapes.

It wasn't until the late twentieth century that biogeomorphology began attracting the attention of more than a handful of researchers.

Research Approaches

There are two approaches to research in biogeomorphology. One is through the statistical and empirically derived means. This is an approach commonly used in the fields of ecology and biology. The approach is simply to employ large replication studies and deriving patterns from statistical data. Whereas taking a more geomorphic research approach tends to derive patterns via theoretic knowledge and detailed measurements of multiple factors. In turn, this uses smaller sample sizes than that of large replication studies.

Biogeomorphological Processes

There are several biogeomorphological processes. Bioerosion is the weathering and removal of abiotic material via organic processes. This can either be passive or active. Moreover, bioerosion is the chemical and or the mechanical weathering of landforms due to organic means. Bioprotection is essentially the effect that organisms have on reducing the action of geomorphic processes. Best shown by algae covering a rock surface acts as a buffer from the waves erosive work. Bioconstruction refers to bioconstructors or ecosystem engineers. Ecosystem engineers are organisms that build mounds, dams, reefs etc. More specifically they are organisms that change the environment physically, directly or indirectly controlling the abundance of resources available in the environment accessible to organisms.

Themes of Complex Systems in Biogeomorphology

There are four main themes that underline the complex systems within biogeomorphology. The first of which is multiple causality. Multiple casualty is the way in which biota is deposited. More specifically, multiple casualty is caused by various processes.

That is, processes such as fires, floods and hillslope instability directly or indirectly determining the distribution of flora and in turn fauna. Ecosystem engineers are another theme underlying complex system of biogeomorphology. These organisms have the most profound effect on the overall ecosystem structure. Some of the most common ecosystem engineers are earthworms. Earthworms aid in the production of humus and increase both soil aeration and area for roots and root hairs to utilize. With more space for roots, this can increase soil stability. Another strong example of ecosystem engineers are beavers. Beavers can increase the sedimentation in a channel as well as increase runoff rates due to a reduction of vegetative cover needed to build their dams. Ecological topology is another theme of complex systems in biogeomorphology. This theme focuses on how the biota varies based on geographic location. This ecological topology is controlled by a concept called stability domain. Stability domain describes the interaction of a set species and certain abiotic factors that act as a medium to the function and structure of an environment. The final of the four underlying themes of the complex systems in biogeomorphology is ecological memory. Ecological memory is where certain biotic and abiotic factors have a recursive relationship and therefore can be encoded in organisms and the immediate environment. An example of this can simply be the flame retardant properties in the bark of Coastal Sequoias due to the recurrence of fires.

Climate Change

Biogeomorphology and ecogeomorphology can aid with assessing the impacts of global climate change. This can especially be seen in coastal and estuarine systems due to; sea level rise, increased global temperatures, increased sea temperature, a higher frequency in and intensity of storms, and varying distributions of precipitation. Biogeomorphology can outline some of the effects of climate change due to biocomplexity. Biocomplexity is in reference to the complex way in which organisms interact with their environment and its effect on biodiversity. Using statistical data, one can derive how these environmental changes will affect the biodiversity of different trophic levels and different keystone species.

Coastal Biogeomorphology

Since the 1990s, biogeomorphology has developed as an established research field examining the interrelationship between organisms and geomorphic processes in a variety of environments, both marine, and terrestrial. Coastal biogeomorphology looks at the interaction between marine organisms and coastal geomorphic processes. Biogeomorphology is a subdisclipline of geomorphology.

This can include not only microorganisms and plants, but animals as well. These interactions are important factors in the development of certain environments like salt marsh, mangrove and other types of coastal wetlands as well as influencing coastal and shoreline stability.

The shape of coastlines can be influenced by biological processes.

Main Processes

There are three main processes related to biogeomorphology: bioerosion, bioprotection, and bioconstruction. Bioerosion is the erosion of ocean substrates by living organisms. Bioprotection refers to the protection of substrate from various forms erosion by the presence of organisms, and the structures they create (i.e. coral reefs). Finally bioconstruction refers to the physical construction of biological structures on ocean substrate. Marine biota interact with landform processes by building structures, accumulating carbonate sediments, accelerating erosion by boring or bioturbation, and marine plant life contribute to shoreline stability, especially in marsh and wetland environments.

Role in Shoreline Stability

The interaction between marine biota and geologic processes is important to shoreline stability, especially in soft sedimentary environments where sediments are more likely to erode away. Benthic and planktonic organisms, as well as shellfish filter, package, and even bind fine sediments together in tidal regions. This action reduces turbidity in the area by solidifying and protecting loose, soft sediments, and thus allowing more colonization by other organisms. If disturbance of these soft sediments occurs, particularly through human interaction such as shellfish harvesting, dredging, or the introduction of toxins, the environment may drastically change. If this occurs, and marine biota are removed from the environment, erosion can occur, or increase, especially in regions prone to wave action and tidal re-suspension.

Phytogeomorphology

Phytogeomorphology is the study of how terrain features affect plant growth. It was the subject of a treatise by Howard and Mitchell in 1985, who were considering the growth and varietal temporal and spatial variability found in forests, but recognized that their work also had application to farming, and the relatively new science (at that time) of precision agriculture. The premise of Howard and Mitchell is that landforms, or features of the land's 3D topography significantly affect how and where plants (or trees in their case) grow. Since that time, the ability to map and classify landform shapes and features has increased greatly. The advent of GPS has made it possible to map almost any variable one might wish to measure. Thus, a very increased awareness of the spatial variability of the environment that plants grow in has arisen. The development of technology like airborne LiDAR has enabled the detailed measurement of landform features to better than sub-meter, and when combined with RTK-GPS (accuracies to 1mm) enables the creation of very accurate maps of where these features are. Comparison of these landform maps with mapping of variables related to crop or plant growth show a strong correlation.

Phytogeomorphology and Precision Agriculture

While phytogeomorphology studies the relationship between plants and terrain attributes in general, it can also apply to precision agriculture by studying crop growth temporal and spatial variability within farm fields. There is already a volume of work, although they don't use the term phytogeomorphology specifically, that considers farm field terrain attributes as affecting crop yield and growth, Moore et al. provide an early overview of the application of terrain features to precision agriculture, but one of the earliest references to this phenomenon in farming is that of Whittaker in 1967. More recent work includes a six-year study of temporal and spatial yield stability over 11 years and a detailed study of the same on a small patch farm in Portugal (and references therein). This variability can be exploited to produce higher yields and reduce the environmental impact of farming - consequently returning a higher profit to the farmer in terms of higher overall yields and lesser amounts of inputs. The new science of Sustainable Intensification of Agriculture which is addressing the need for higher yields from existing fields can be fulfilled by some of the practical applications of phytogeomorphology applied to precision agriculture.

Work in this area has been happening for some years, but it is slow and sometimes tedious work that necessarily involves multiple years of data, very specialized software tools, and long compute times to produce the resulting maps.

Phytogeomorphologically Defined Management Zones

Typically, the objective of precision agriculture is to divide the farm field into distinct management zones based on yield performance at each point in the field. 'Variable rate technology' is a relatively new term in farming technology that refers to spreaders, seeders, sprayers, etc. that are able to adjust their rates of flow on the fly. The idea is to create a 'recipe map' for variable rate farm machinery to deliver the exact quantity of amendments required at that location (within that zone of the field). The literature is divided on how to properly define management zones.

In the geomorphological approach to defining management zones it is found that topography aids in at least partially defining how much yield comes from which part of the field. This is true in fields where there are permanently limiting characteristics to parts of the field, but not true in fields where the growth potential is the same all over the field. It can be shown that an index map of yield (shows areas of consistent over-performance of yield and areas of consistent under-performance) correlates well with a landform classification map. Landforms can be classified a number of ways, but the simplest to use software tool is LandMapR. An early version of the LandMapR software is available through the Opengeomorphometry project hosted under the Google Code project.

Climatic Geomorphology

Atolls like Atafu in Tokelau in the Pacific Ocean are landforms associated to tropical climate. No atoll exists outside the tropics.

Climatic geomorphology is the study of the role of climate in shaping landforms and the earth-surface processes. An approach used in climatic geomorphology is to study relict landforms to infer ancient climates. Being often concerned about past climates climatic geomorphology considered sometimes to be an aspect of historical geology. Since landscape features in one region might have evolved under climates different from those of the present, studying climatically disparate regions might help understand present-day landscapes. For example, Julius Büdel studied both cold-climate processes in Svalbard and weathering processes in tropical India to understand the origin of the relief of Central Europe, which he argued was a palimpsest of landforms formed at different times and under different climates.

During the age of New Imperialism in the late 19th century European explorers and scientists traveled across the globe bringing descriptions of landscapes and landforms. As geographical knowledge increased over time these observations were systematized in a search for regional patterns. Climate emerged thus as prime factor for explaining landform distribution at a grand scale. The rise of climatic geomorphology was foreshadowed by the work of Wladimir Köppen, Vasily Dokuchaev and Andreas Schimper. William Morris Davis, the leading geomorphologist of his time, recognized the role of climate by complementing his "normal" temperate climate cycle of erosion with arid and glacial ones. Nevertheless, interest in climatic geomorphology was also a reaction against Davisian geomorphology that was by the mid-20th century considered both un-innovative and dubious. Early climatic geomorphology developed primarily in continental Europe while in the English-speaking world the tendency was not explicit until L.C. Peltier's 1950 publication on a periglacial cycle of erosion.

Climatic geomorphology was criticized in a 1969 review article by process geomorphologist D.R. Stoddart. The criticism by Stoddart proved "devastating" sparking a decline in the popularity of climatic geomorphology in the late 20th century. Stoddart criticized climatic geomorphology for applying supposedly "trivial" methodologies in establishing landform differences between morphoclimatic zones, being linked to Davisian geomorphology and by allegedly neglecting the fact that physical laws governing processes are the same across the globe. In addition some conceptions of climatic geomorphology, like that which holds that chemical weathering is more rapid in tropical climates than in cold climates proved to not be straightforwardly true.

Despite having diminished in importance, climatic geomorphology continues to exist as a field of study producing relevant research. More recently, concerns over global warming have led to a renewed interest in the field. The various subbranches of climatic geomorphology focus on specific climatic environments.

Desert Geomorphology

Desert geomorphology or the geomorphology of arid and semi-arid lands shares many landsforms and processes with more humid regions. One distinctive feature is the sparse or lacking vegetation cover, which influences fluvial and slope processes, related to wind and salt activity. Early work on desert geomorphology was done by Western explorers of the colonies of their respective countries in Africa (French West Africa, German South West Africa, Western Egypt), in frontier regions of their own countries (American West, Australian Outback) or in the deserts of foreign countries such as the Ottoman Empire, the Russian Empire and China. Since the 1970s desert geomorphology in Earth has served to find analogues to Martian landscapes.

Periglacial Geomorphology

As a discipline periglacial geomorphology is close but different to Quaternary science and geocryology. Periglacial geomorphology is concerned with non-glacial cold-climate landforms in areas with and without permafrost. Albeit the definition of what a periglacial zone is not clear-cut a conservative estimate is that a quarter of Earth's land surface has periglacial conditions. Beyond this quarter an additional quarter or fifth or Earth's land surface had periglacial conditions at some time during the Pleistocene. In periglacial geomorphology noted researchers include Johan Gunnar Andersson, Walery Łoziński, Anders Rapp and Jean Tricart.

Tropical Geomorphology

If the tropics is defined as the area between 35° N and 35° S, then about 60% of Earth's surface lies within this zone. During most of the 20th century tropical geomorphology was neglected due to a bias towards temperate climates, and when dealt with it was highlighted as "exotic". Tropical geomorphology do mainly differ from other areas in the intensities and rates at which surface processes operate, and not by the type of processes. The tropics are characterized by particular climates, that may be dry or humid. Relative to temperate zones the tropics contain areas of high temperatures, high rainfall intensities and high evapotranspiration all of which are climatic features relevant for surface processes. Another characteristic, that is not related to present-day climate per see, is that a large portion of the tropics have a low relief which was inherited from the continent of Gondwana. Julius Büdel, Pierre Birot and Jean Tricart have suggested that tropical rivers are dominated by fine-grained suspended load derived from advanced chemical weathering, and this would make them less erosive than rivers elsewhere. Some landforms previously thought as typically tropical like bornhardts are more related to lithology and rock structure than climate.

Morphoclimatic Zones

Yardangs in Lut Desert, Kerman Province. Deserts constitute undisputed morphoclimatic zones.

Climatic geomorphologists have devised various schemes that divide Earth's surface into various morphoclimatic zones; that is, zones where landforms are associated to present or past climates. However, only some processes and landforms can be associated with particular climates, meaning that they are zonal; processes and landforms not associated with particular climates are labelled azonal. Despite this, azonal processes and landforms might still take on particular characteristics when developing under the influence of particular climates. When identified, morphoclimatic zones do usually lack sharp boundaries and tend to grade from one type to another resulting in that only the core of the zone has all expected attributes. Influential morphoclimatic zoning schemes are those of Julius Büdel and of Jean Tricart and André Cailleux. Büdel's schemes stresses planation and valley-cutting in relation to climate, arguing the valley-cutting is dominant in subpolar regions while planation is so in the tropics. As such this scheme is concerned not only with processes but also with end-products of geomorphic activity. The scheme of Tricart and Cailleux emphasizes the relationship between geomorphology, climate and vegetation. An early attempt at morphoclimatic zoning is that of Albrecht Penck in 1910, who divided Earth in three zones depending on the evaporation-precipitation ratios.

A 1994 review argues that only the concepts of desert, glacial, periglacial and a few coastal morphoclimatic zones are justified. These zones amounts to about half of Earth's land surface, the remaining half cannot be explained in simple terms by climate-landform interactions. The limitations of morphoclimatic zoning were already discussed by Siegfried Passarge in 1926 who considered vegetation and the extent of weathered material as having more direct impact than climate in many parts of the World. According to M.A. Summerfield large-scale zoning of the relief of Earth's surface is better explained on the basis of plate tectonics than on climate. An example of this are the Scandinavian Mountains whose plateau areas and valleys relate to the history of uplift and not to climate.

Piotr Migoń has questioned the validity of certain morphoclimatic zonation schemes since they are named after processes, like planation, that might not occurring at all in large swathes of the zone. Referring to the 1977 scheme of Büdel Migoń states.

Is it really helpful to have the Volcanic Cordillera of Mexico, coastal ranges of southeast Brazil, plains of East Africa, the escarpments of Western Ghats and the mountains of Taiwan in the same zone, labelled as the 'peritropical zone of excessive planation'.

Hydrogeomorphology

Hydrogeomorphology has been defined as "an interdisciplinary science that focuses on the interaction and linkage of hydrologic processes with landforms or earth materials and the interaction of geomorphic processes with surface and subsurface water in temporal and spatial dimensions." The term 'hydro-geomorphology' designates the study of landforms caused by the action of water. By this definition hydro-geomorphology is inseparable part of geomorphology moreover fluvial geomorphology, because water is one of the most important agents in forming and shaping of landforms. From the groundwater point of view integration of geological, structural and hydrological data with hydro-geomorphologic data is very much useful in finding out the groundwater potential zones with fruitful results. The science relating to the geographical, geological, and hydrological aspects of water bodies and to changes to these aspects in response to low variations and to natural and human caused events, such as heavy rainfall or channel straightening is the hydro-geomorphology.

Hydrogeomorphology describes and evaluates the environment, in which water circulates, thus providing the information to understand the situation and to make the proper decisions. Quantitative study of drainage basin provides the theoretical base for the hydrogeomorphic approach, suggesting that certain unvarying drainage basin characters can be correlated to the hydrologic response of a basin. The measurable description of a drainage basin can be grouped in to linear aspect of channel network, areal aspect of drainage basin, relief aspect of channel system and basin form. Hydro-geomorphology is science that deals with occurrences of water with respect to landform. Hydrogeomorphology of a drainage basin is a function of rainfall kinematics, surface topography, drainage basin morphology and runoff etc. All these aspects are regarded as the potential to describe hydrogeomorphic properties of the drainage basin. Because of non-availability of hydrological data, discharge data and sediment load data over a sufficient period of time in ungauged catchments, various investigators have used the drainage basin parameters to study the hydrogeomorphology of the drainage basin. Hydrogeomorphic studies of drainage basin often suffer a setback due to lack of long-term data. Therefore, there is need to extrapolate the results of few small subsystems to other hydrologically and geomorphologically similar basins, which mostly remain ungauged for want of enormous resource and time involved in instrumentation and monitoring them. Importance of hydrology for geomorphological purposes has been increasingly appreciated among geomorphologists in the last few decades. Earlier geomorphologists were bounded to use different unconventional approach to evaluate the characteristics of rivers and drainage basins to get proper idea of various aspects of the water crisis.

Feedback effect in hydrogeomorphology. This river in Addilal is incising, due to reforestation on the upper slopes, hence much less sediment is brought in and the river erodes its own bed - this is called "clear water effect".

R.E.Horton pioneered the hydrologic and hydromorphometric analysis of basin and provided a rational and systematic base. He framed the geomorphic parameters with hydrologic parameters of the drainage system. Thus hydrological criteria not only assist the geomorphologists to evaluate the hydrogeomorphic characteristics of a drainage basin also facilitate their extrapolation in space and time.

Applied Geomorphology

Applied geomorphology is a field of science where the research outcomes provide information geomorphic landforms or processes that may be of concern to society, and, where relevant, provides solutions to problems of geomorphic context. Applied geomorphology examines geomorphic impacts that affect society, as well as society's impact on geomorphic forms and processes. The application of geomorphic knowledge to problem-solving spans all of the classic traditions of geomorphology and is useful for human–environmental solutions across a broad range of geographies, including coastal shoreline erosion control, landslide risk assessments, and dam-related river management issues. Resources, particularly maps, models, and prediction tools, provided by applied geomorphologists are useful to scientists, engineers, consultants, and decision-makers involved with hazards, land-use planning, natural resources, environmental management, and global environmental change. Applied geomorphologists working independently or serving on multidisciplinary advisory panels are well positioned to influence public policy to the benefit of society and the earth sciences.

An understanding of landforms may be of great use, directly or indirectly, to human beings who are influenced by and, in turn, influence the surface features of the earth which they inhabit.

If landforms are properly interpreted, they throw light upon the geologic history, structure and litho logy of a region. According to D.K.C. Jones, applied geomorphology could be defined as "the application of geomorphic understanding to the analysis and solution of problems concerning land occupancy, resource exploitation, and environmental management and planning".

Indeed, all geomorphological knowledge tends to be applied, according to R.G. Craig and J.L. Craft. As each advance in knowledge provides a clear view of how the earth works, geomorphologists can make use of the knowledge for evaluating resources, development projects, locating natural hazards and mitigating the effect of natural disasters.

Geomorphic knowledge and techniques may be applied in the following areas:

- Studying the impact of geomorphic/environmental processes on human society and activities and dealing with problems arising out of such impact.
- Investigating the changes brought about in the geomorphic/environmental processes by human activities and dealing with the problems arising out of such interaction.
- Managing resources and monitoring changes in the geomorphic system to suggest suitable remedial measures for maintaining development at a sustainable level.

Two Main Lines of Application

The application of geomorphology, according to Charley, Schumn and Sugden, may be considered along two lines:

- Geomorphology can be an aid to resource evaluation, engineering construction and planning. In this category we may put resource inventories, environmental management, soil and land evaluation, production of maps for hydrological, erosional and stability control, geomorphic mapping, mapping for land systems and evaluating terrain, classification and retrieval of information on terrain and other matters of use to earth scientists, engineers and planners.

Applied geomorphology in this aspect can be of use in urban planning in different geomorphic environments and in preparation of natural hazard maps, morpho-agricultural regionalisation, land use planning, construction and management of roads.

- Applied geomorphology is also concerned with human beings as geomorphic agents, in terms of their planned or inadvertent effects on geomorphic processes and forms.

Human beings have over time tried to tame and modify geomorphic/environmental processes to suit their economic needs. Embankments have been built to check flooding of rivers; meandering courses of rivers have been straightened and channels diverted; coastal areas have been sought to be protected against wave erosion by building walls; there have been attempts to stabilise sandy areas through plantation, and check soil erosion through afforestation. These are some examples of planned activities by human beings that have an impact on geomorphic forms and processes.

The inadvertent effects of human activities on geomorphic forms and processes are many: Forests are cleared and grasslands burnt for cultivating crops or for building settlements; mining activities and water extraction cause subsidence of land; building and mining activities result in modification of terrain; excessive, unplanned deforestation causes accelerated soil erosion and increase in sediment load leading, in turn, to recurrent floods and riparian decay. Pollution has been a major inadvertent effect of human economic activity. Dams cause changes in river load and accelerated erosion. High altitude construction has modified permafrost.

Specific Applications

We consider here some of the applications of geomorphology to the types of problems commonly encountered by geologists, engineers and planners.

Geomorphology and Hydrology

Water used by human beings is available from different sources—streams, lakes and rivers on the surface of the earth or groundwater. Different stratigraphic and lithological zones present different conditions of surface and groundwater.

Limestone terrains vary widely and the ability to yield water depends on the type of rock. Permeability in limestones may be primary or secondary. Primary permeability depends upon the presence of initial interconnecting voids in the calcerous sediments from which the rock was formed. Secondary (or acquired) permeability occurs because of earth movements such as faulting, folding, warping, and due to solution or corrosion mechanism.

This secondary permeability varies notably with respect to the topography of a region, being greatest beneath and adjacent to topographic lows or valleys. Much of the groundwater in karst terrain is confined to solution channels.

In early stages of karst evolution conditions are not too different from those of other types of landscapes with similar relief. But as the cycle advances, a large proportion of water is diverted to solutionally opened passageways, and surface water gets diminished. The main source of water in such regions then are karst springs. Such springs may supply water to meet moderate demands, but the quality of water may be affected by pollutants and bacteria.

The sources of the spring water should be determined in such a case of pollution. The swallow holes and sinkholes feeding water to the underground drainage systems emerging as springs may be located. This can be done by putting some colouring material, such as fluorescein, into the water entering nearby swallow holes (or sinkholes) and testing the various spring waters to find out their source. A knowledge of the structural geology of the region is of use in this context, as groundwater moves down rather than up the regional dip.

The ease with which water may be obtained in a limestone region depends on the geomorphology of the area. If the limestones have enough permeability and are capped by sandstone layer, there may be no difficulty in obtaining wells of large yields. Moreover, the water would get naturally filtered as it passes through the sandstone beds.

If, however, the limestone is dense and compact, with little mass permeability, movement of groundwater will be largely through secondary openings. In such circumstances, the yield of water may be low or, even if adequate, subject to contamination. Karst plains lack a filtering cover and sinkholes, swallow holes or karst valleys within an area of clastic rocks should cast doubt on the purity of the water from springs nearby.

Groundwater potential in glaciated regions can be determined on the basis of understanding the geomorphic history of the area, characteristics of glacial deposits and landform. Outwash plains, valley trains and intertill gravels are likely to yield large volumes of water. Most tills are poor sources of water because of the clay in them, but they contain local strata of sand and gravel which may hold and supply enough water for domestic needs.

Buried preglacial and interglacial valleys could be good sources of groundwater. Their presence (or absence) may be detected by studying the preglacial topography and geomorphic history of the area. Buried valleys are located by constructing bedrock topography maps of glaciated areas.

Geomorphology and Mineral Exploration

Mineral deposits are associated with geological structure. Landscape characteristics of the specific localities could indicate such geological structures.

Surface Expression of Ore Bodies

Some ore bodies have obvious surface expressions as topographic forms, as outcrops of ore, gossan, or residual minerals, or structural features such as faults, fractures and zones of breccia. Lead-zinc lodes could be marked by a conspicuous ridge as in the case of Broken Hill, Australia. Quartz veins could stand out prominently as they are much more resistant to erosion than the unsilicified surroundings, as in Chihuahua, Mexico. Some veins (calcite, for instance) and mineralised areas may be indicated by depressions or subsidence features.

Weathering Residues

Many economically important minerals are the weathering residues of present or ancient geomorphic cycles and geomorphology can be of use in searching for such minerals. Iron ore, clay minerals, caliche, bauxite and some ores of manganese and nickel may be such weathering residues. Weathering and erosion are constantly at work on the rocks of earth's surface, and the products of rock weathering may be of economical value.

The surfaces on which residual weathering products commonly form are pleneplain or near-pleneplain surfaces. Such minerals are more commonly to be found upon remnants of Tertiary erosional surfaces above present base levels of erosion. Bauxite, for instance, is either the residue of a small amount of insoluble aluminous material in dolomites and limestones or it is the direct product of the weathering of aluminous minerals.

Placer Deposits

Placer deposits are mixtures of heavy metals which are aggregates of materials derived through chemical weathering or erosion of metallic formation. Placer concentration of minerals results from definite geomorphic processes and, found in specific topographical positions, may have a distinctive topographic expression. The type of rock forming the bedrock floor may influence the deposition of placers.

Residual placers or 'seam diggings' are residues from the weathering of quartz stringers or veins, are usually of limited amount, and grade down into lodes. Colluvial placers are produced by creep downslope of residual materials and are thus transitional between residual placers and alluvial placers.

Gold placers of this type have been found in California, Australia, New Zealand, and elsewhere. Part of the tin placers of Malaya is colluvial placers (the koelits) and parts are alluvial placers (the kaksas). About one-third of the world's platinum is obtained from alluvial placers in Russia, Colombia, and elsewhere. Gold, tin and diamonds are among the more important minerals obtained from alluvial placers.

Diamonds in the Vaal and Orange River districts of South Africa, the Lichtenburg area of South Africa, the Belgian Congo, and Minas Geraes, Brazil, are obtained from alluvial placers. About 20 per cent of the world's diamonds comes from placer deposits. Aeolian placers have yielded gold in Australia and Lower California, Mexico. Bajada placers form in the gravel mantle of a pediment and in the confluent alluvial fans of a bajada.

They are more likely to be found near a mountain base than out on the more gentle slopes of a basin fill. Beach placers have yielded gold in California and Alaska, diamonds in the Namaqualand district of South Africa, zircon in India, Brazil, and Australia, and ilmenite and monazite in Travancore, India.

Location of placers may be aided by drilling and geophysical testing. A magnetic survey will usually be helpful because magnetite is likely to be associated with gold. If the bedrock is a basic type with a higher magnetic intensity than the placer gravels, areas of magnetic "lows" may reflect the positions of the filled channels.

"Knowledge of the bedrock geology, application of geophysical surveying, test drilling and aerial- photograph interpretation all contribute their parts to exploration for these buried placers, but most fundamental to this search is a thorough understanding of the geomorphic history of the region," observes W.D. Thornbury.

Oil Exploration

Many oil fields have been discovered because of their striking topographic expression. Mineral oil is considered to have been formed by the decay and decomposition of organic matter. After formation, this oil gets trapped in rocks under structural traps or stratigraphic traps. Sedimentary strata are folded into anticlines and synclines allowing the permeable and impermeable strata to get closer, and the mineral oil are well-preserved within the upper permeable and the lower impermeable beds.

Generally mineral oil is found in the porous and permeable rock structures with lower layers of impermeable rocks. Sandstone and limestone provide ideal locations of mineral oil as they are porous and permeable. The shale below acts as the impermeable bed. In regions of heavy tropical forests, where topography cannot be seen through the forest, tonal differences may indicate an anticlinal or domal structure.

More subtle evidence of geologic structures favourable to oil accumulation is being made use of today in the search for oil. Drainage analysis of a terrain shown on aerial photography is one such technique. A sophisticated perception of drainage anomalies of an area is required, and a geomorphologist would most likely possess the requisite knowledge. Drainage analysis is particularly useful in regions where rocks have low dips and the topographic relief is slight.

According to Leverson, many oil and gas sources are associated with unconformities—ancient erosion surfaces; hence a petroleum geologist must deal with buried landscapes. Where ancient erosion surfaces shorten permeable beds and are later sealed over by deposits, the erosion surfaces become stratigraphic traps, most of which are along unconformities.

Geomorphology and Engineering Works

Engineering works most often involve evaluation of geologic factors of one type or another; terrain characteristics are among the most common factors.

Road Construction

The most feasible highway routes would be best determined by the topographic features of the area. Knowledge of the geologic structure, lithological and stratigraphic characteristics, strength of the surficial deposits, geomorphic history of the area, among other things are of importance in road engineering.

A route over a karst plain necessitates repeated cut and fill otherwise the road will be flooded after heavy rains as sinkholes fill with surface runoff. Bridge abutments in a karst region should be so designed that they will not be weakened by enlarged solutional cavities which are likely to be present.

Glacial terrains present many types of engineering problems. A flat till plain is topographically ideal for road construction, but in areas where end-moraines, eskers, kames or drumlins exist there is need for cut and fill to avoid circuitous routes. Muck areas, which mark sites of former lakes, are unsuited for roads which are to carry heavy traffic. If a road is built across them as they are, heavy traffic will cause the plastic materials beneath the lake floor to flow, and 'sinks' in the road bed will result. To avoid this, the lacustrine fill may have to be excavated and replaced with materials that will not flow under heavy load.

Areas with the considerable relief which characterises late youth and early maturity will necessitate much bridge construction and many cuts and fills. In such areas landslides, earth flows and slumping become serious problems.

In highway construction designed to carry heavy traffic, the nature of the soil beneath a road surface, or what is called the subgrade, has become increasingly significant because of its control over the drainage beneath a highway. The lifetime of a highway, under moderate loads, is determined largely by two factors: the quality of the aggregate used in the highway and the soil texture and drainage of its subgrade.

Thus an appreciation of the relationships of soils to varying topographic conditions and type of parent material becomes essential in modern highway construction. A knowledge of soil profiles, which to a large degree reflect the influence of geomorphic conditions and history, is basic. Poor highway performance characterises silty-clay subgrades with a high water table, and best performance is found on granular materials with a low water table.

Dam Site Selection

Selecting sites for the construction of dams would be greatly helped by a synthesis of knowledge concerning geomorphology, lithology, and geologic structure of terrains.

Five main requirements of good reservoir sites depend on geologic conditions, according to Kirk Bryan:

(1) A water-tight basin of adequate size; (2) a narrow outlet of the basin with a foundation that will permit economical construction of a dam; (3) opportunity to build an adequate and safe spillway to carry surplus waters; (4) availability of materials needed for dam construction (this is particularly true of earthen dams); and (5) assurance that the life of the reservoir will not be too short as a result of excessive deposition of mud and silt.

Limestone terrain, for instance, may prove a difficult one for constructing a dam. The bedrock surface may be irregular because of differential solution and, unless a true picture of the subsurface is understood, it could lead to avoidable expenditure.

A construction in a valley is desirable from the standpoint of the size of the dam that will have to be built, but it may not always be a good dam site. In glaciated areas, where buried bedrock valleys containing sand and gravel fills are common, surficial topography may not give an adequate picture of sub-surface conditions.

On a superficial level, a dam could be situated at a constriction in a valley between a valley wall on one side and a spur end on the other. However, it is possible that the subsurface topography is not suitable; there may be a buried preglacial valley with sand and gravel in it. The dam, if built on such a site, would leak.

Construction of Air Strips while selecting sites for air strips, where aeroplanes may land and take off from, requires engineering skill, the engineer could well benefit from a geomorphologist's knowledge of landscape characteristics. The best sites for air strips would show an extensive flat surface with resistant geomaterials available: here a safe and suitable runway can be constructed. Furthermore, the ground surface should show an almost level slope, the area should be free of floods and marked by good visibility (may be free of frequent and intense fog). These characteristics could be gathered from a morphological map of the region where the air strip is to be built.

Locating Sand and Gravel Pits

Sand and gravel have many engineering as well as commercial and industrial uses. Selection of suitable sites for sand and gravel pits will entail evaluation of such geologic factors as variation in grade sizes, lithologic composition, degree of weathering, amount of overburden and continuity of the deposits. Sand and gravel may be found as floodplain, river terrace, alluvial fan and cone, talus, wind-blown, residual and glacial deposits of various types.

All have distinctive topographic relationships and expressions and varying inherent qualities and possibilities of development. Recognition of the type of deposit is essential to proper evaluation of its potentialities. Demand for gravel is generally greater than for sand, particularly in recent years with the decreased use of plaster in home construction, and thus knowledge of the percentages of various grade sizes is important.

Floodplain deposits are likely to contain high proportions of silt and sand and show many variable and heterogeneous lateral and vertical gradations. Alluvial fan and cone gravels are angular in shape as well as variable in size, especially near their apices. Talus materials, in addition to being angular, are too large to be suitable for most uses and are limited in extent.

Wind-blown sands may be satisfactory sources of sand but have no gravel in them. Residual deposits lack assortment and are likely to contain pebbles that are too deeply weathered to be suitable for use as aggregate in cement work. A high percentage of iron-coated chert when used as aggregate usually has deleterious effects. Residual deposits are furthermore likely to be limited in extent.

Terraced valley trains and outwash plains are usually favourable sites for pits, for they do not have a thick overburden and are usually extensive.

Geomorphology and Military Geology

Thornbury points out the importance of geomorphology in the context of war.

World War I was largely stabilised trench warfare, and the information that was most useful was more geologic than geomorphic in nature. Information about the kind of rock that would be encountered in digging trenches, in mining and countermining and the possibilities of water supply and supplies of other geologic materials was most utilised. Topography did play a role in maneuvering and planning routes of attack, but it can hardly be said that the Allies utilised basic geomorphic knowledge to any great extent.

"With development of the blitzkrieg type of warfare during World War II, topography became more important, because the effectiveness of a blitz depends to a large degree upon the trafficability of the terrain. As a consequence, in more recent years terrain appreciation or terrain analysis have become semi-magic words with the military. A geomorphologist may lack knowledge as to how best to utilise terrains in military operations, but certainly his concepts of terrain conditions are far more adequate than those of the military specialist or other geologists for that matter. He appreciates that landforms are the result of an interaction of geomorphic and geologic processes through time; that landforms are not groups of unrelated and haphazard individual forms but have systematic relationships that reveal their origins and tell much about the underlying bedrock geology and structure, as well as the soils and vegetation of a region. As Erdmann put it, the geomorphologist has 'an eye for the ground or an instinctive eye for configuration, the judgement of how distant ground, seen or unseen, is likely to lie when you come to it. Terrain is the common denominator of geology and war.' Whether it is in connection with the interpretation of topographic maps or aerial photographs, this basic appreciation of different types of terrain is fundamental to a proper planning of military campaigns."

Regarding the use of aerial photographs in this connection, Hunt stated: "Even where geologic maps are lacking or are on such a small- scale as to be practically useless for tactical intelligence, geologic principles can be applied with advantage to interpreting the terrain from aerial photographs. Little training in reading vertical photographs is required to recognise mountains, hills, lakes, rivers, woods, plains or some kinds of swamps. But much more than that can and should be interpreted from the pictures for the purposes of acquiring complete terrain intelligence. It is essential to know the kind of hill, the kind of plain, the kind of river or lake, and so on, because by knowing this it is frequently possible to reconstruct the geology. The interpreter, with some confidence, can then make predictions as to water supply, the kind and depth of soil, traffic-ability, ground drainage and other construction problems, construction materials, movement and cover, and many of the other elements that are essential to an adequate estimate of the terrain situation. In brief, therefore, aerial photographs are useful to the preparation of terrain intelligence insofar as they provide information on the geology of' the area. Identification of a hill or other terrain feature is but a small part of the story that can be read from a photograph; all important is the recognition of the significance of the particular landform, in terms of kind of ground and slope."

Geomorphology and Urbanisation

Geomorphological knowledge applied to urban development has become important enough to grow into a separate branch, namely, urban geomorphology. This branch of geomorphology is concerned with "the study of landforms and their related processes, materials and hazards, ways that are beneficial to planning, development and management of urbanised areas where urban growth is expected," according to R.U. Cooke.

A city or town depends for its stability, safety, basic needs and, later, its expansion on geomorphological features: lithological and topographical features, hydrological conditions and geomorphic features. An urban geomorphologist begins work even before urban development through field survey, terrain classification, identification and selection of alternative sites for settlements. During and after urban development, an urban geomorphologist would be concerned with studying the impact of natural events on' the urban community and that of urban development on the environment.

It has been pointed out by R.U. Cooke that "various geomorphological problems hitherto not understood by the planners and engineers lead to destruction and damage to urban settlements in varying environmental realms viz., settling of foundation materials in the dry lands of oil-rich states and in the periglacial regions; destruction of foundations by weathering processes; damage of highways; damage to buildings through inundation during floods in the subtropical humid regions etc.

All these and many other problems arise in part from mismanagement or misunderstanding of geomorphological conditions." Very little attention is paid to understanding the geomorphological conditions before the development of existing urban centres mainly in the developing countries. This results in uncontrolled growth, giving rise to squatter settlements or shanty towns. It commonly creates serious social and environmental problems.

Geomorphology and Hazard Management

Events, natural or man-induced, exceeding a tolerable level or of an unexpected nature may be called hazards. A geomorphic hazard, says Chorley, may be defined as "any change, natural or man- made, that may affect the geomorphic stability of a landform to the adversity of' living things". These hazards may arise from long-term factors such as faulting, folding, warping, uplifting, subsidence caused by earth movements, or changes in vegetation cover and hydrologic regime caused by climatic change. More immediate and sudden hazards are volcanic eruptions, earthquakes, landslides, avalanches, floods, etc.

Geomorphic knowledge can be of use in identifying and predicting such hazards and in assessing their effects and proper management.

Regular measurement of seismic events and earth tremors by seismic methods; regular measurement of ground surface, mainly tilt measurement by tilt metres; constant measurement of temperature of crater lakes, hot springs, geysers, fumaroles; monitoring of gases coming out of craters, hot springs, geysers; monitoring of changes in the configuration of dormant or extinct volcanoes by lasers; measurement of local gravity and magnetic fields and their trends etc., help in making predictions of possible eruptions in the areas having past case histories of vulcanism. The path of lava flow can be better predicted on the basis of detailed analysis of topography and identification of possible eruption points.

The geomorphic knowledge of the behaviour of a river system and its morphological characteristics viz. channel geometry, channel morphology and channel pattern, river metamorphosis, bank morphology etc., may help in controlling river floods through flood control measures.

These include steps:

- To delay the return of runoff resulting from torrential rainfall to the rivers.
- To hasten the discharge of water (by straightening the meandering channels).
- To divert the flow of water (through, diversion channels).
- To reduce the impact of floods (through construction of protective embankments).
- To forewarn the occurrence of floods.

Without the knowledge of the nature of erosion in the upper catchment area and sediment load characteristics of the river, the construction of levees to confine the flood water within the valley may prove disastrous: if the rate of erosion is very high in the upper catchment area, resulting in high sediment load, there would be more sedimentation in the valley causing a gradual rise in the river bed; this may lead to sudden flash floods whenever the levee is breached.

Earthquakes may be natural or man-induced geomorphic hazards. The geomorphic knowledge of the stability of terrain and probable impacts of man-made structures (such as dams and reservoirs) on crustal stability is of paramount importance in identifying weaker zones which are likely to be affected by seismic events. Similarly, the geomorphic study of the nature of hill slopes and their associated lithologies enable us to know the stability or instability of the hill slopes. This knowledge would help in identifying and mapping unstable hill slopes unsuitable for human settlements and road construction.

Geomorphology and Regional Planning

Applied geomorphology has a place in regional planning. A balanced growth of a country's economy requires a careful understanding of what each region offers in terms of resources, natural and human. Detailed information on topography, soils, hydrology, litho-logy and terrain characteristics are of obvious interest to enlightened regional planners who may then devise development projects best suited for the region.

Other Applications

The applications of geomorphic principles are most striking in the fields discussed above. But there are several other areas in which applied geomorphology is of use. As Thornbury points out, soil maps are to a considerable degree topographic maps, and the differentiation of the various members of any soil series rests fundamentally upon the different topographic conditions under which each member of the soil series developed. Modern beach engineering, to be successful, must be based upon an appreciation of the processes of shoreline development.

The problem of soil erosion is essentially a problem involving recognition and proper control of such geomorphic processes as sheet-wash erosion, gulleying, mass-wasting and stream erosion. The severity of erosion is not determined by the angle of slope alone. It may not be serious on steep slopes where those slopes are underlain by permeable materials, and it may be serious on slight slopes where they are on impermeable materials. The related problem of land classification also entails an appreciation of varying types of terrains and the best uses that may be made of them.

Application of geomorphology can be of immense use in controlling the adverse effects of human activities on geomorphic forms and processes. Indeed, this field has also developed into a separate branch, 'anthrop geomorphology'.

Technique of Applied Geomorphology

Applied geomorphology deals with the interactions of geomorphology with anthropogenic activities.

So the basic aspects of applied geomorphology are as follows:

- Mapping of landforms viz. slope elements which affect and modify human activity.
- Attempting to interpret aerial photographs and images taken by remote sensing methods.
- Monitoring the environmental changes, especially when such changes are not sustainable in nature.
- Endeavouring to assess the causes of unsustainable changes.
- Proposing remedies for hazards caused by unsustainable changes.

Interpretation of Aerial Photographs and Satellite Images

Preparation of specialised maps and interpreting them has become easier and accurate with the introduction of air photographs and satellite imageries. Air photographs (taken from aeroplanes) are taken on different scales and the distributional patterns of relevant features are transferred on the maps bearing the same scales to make them up-to-date.

Nowadays, aerial photographs are being used for evaluating landforms and land use vis-a-vis city developmental plans, major construction projects, etc. Satellite imageries are useful for studying global and country-level climatic phenomena (weather forecasting has become more accurate with the introduction of meteorological data gathered by satellites) but these imageries are also of paramount importance in mineral prospecting, preparation of land use inventories and forecasting agricultural output etc.

Remote Sensing

Remote sensing deals with the collection of information regarding objects from a certain distance without coming into contact with them. An assembly of electro-optical devices called 'sensors' as well as cameras measure spectral behaviour of objects under study. Nowadays, the most widely used remote sensing techniques are linked with the sensing of electromagnetic radiation emitted from the terrestrial objects. Different objects have different scattering properties called signatures owing to their different molecular composition. A thorough knowledge of signatures is vital for interpreting satellite images.

Importance of Remote Sensing Technique

Remote sensing is necessary for sustainable management of natural resources like soil, forest, crops, oceans, urban and town planning etc. Resource planners require such techniques for timely information on the condition and extent of resources. Since these resources are dynamic and replenishable in nature, the ground-based monitoring systems can hardly monitor the condition of these resources without a gap of days or weeks. Satellite-based surveys enjoy a definite advantage of repetitivity.

These surveys are conducted from a height of 500-900 km above the surface. "Nowadays Geographical Information Systems or GIS technology has been used along with remote sensing techniques. GIS may be defined as spatial, integrated data-handling programmes used to gather, store and retrieve spatial data from the real world. GIS contain selected data, only those properties geographical investors consider to be relevant."

Remote Sensing Survey enjoys the following advantages over ground surveys:

- Synoptic view or wide coverage of a large area is possible.
- Permanent record of ground conditions is subject to verification later at any time.
- Interpretation of remote sensing data requires much less time than do cumbersome ground surveys.
- Remote sensing technique is capable of accessing thermal and microwave regions not accessible to the naked eye.
- Ground surveys involve more time, money and infrastructure in comparison to remote sensing surveys.
- Ground surveys, if repeated, are highly uneconomical.
- The same remote sensing data is useful for different purposes; for example, the same data may be used by soil scientists for soil surveys, by geohydrologists for groundwater surveys or by agricultural scientists for crop surveys.
- Remote sensing surveys are free from handicaps like bad weather conditions.

References

- What-fluvial-geomorphology-fgm, riversmart-resources: extension.umass.edu, Retrieved 02 June, 2020
- Gutiérrez, Mateo, ed. (2005). "Chapter 1 Climatic geomorphology". Developments in Earth Surface Processes. 8. pp. 3–32. doi:10.1016/S0928-2025(05)80051-3. ISBN 978-0-444-51794-4
- Sidle, Roy C.; Onda, Yuichi (2004). "Hydrogeomorphology: overview of an emerging science". Hydrological Processes. 18 (4): 597. doi:10.1002/hyp.1360
- Applied-geomorphology-meaning-two-main-lines-specific-applications-and-techniques-779: geographynotes.com, Retrieved 28 April, 2020
- Stallins, J. Anthony (2006). "Geomorphology and Ecology: Unifying Themes for Complex Systems in Biogeomorphology". Geomorphology. 77.3 (3–4): 207–216. doi:10.1016/j.geomorph.2006.01.005

Geomorphic Processes

3

- Igneous Process
- Aeolian Process
- Erosion and Tectonics
- Fluvial Processes
- Hillslope Processes and Mass Movement
- Marine Process

Geomorphic processes refer to the physical and chemical processes that modify the surface of the Earth. It includes igneous process, aeolian process, fluvial process, hillslope process, marine process, etc. This chapter has been carefully written to provide an easy understanding of these geomorphic processes.

Igneous Process

Igneous rock is formed when liquid rock freezes into a solid rock. This molten material is called magma when it is in the ground and lava when it is on the surface. Only the Earth's outer core is liquid; the Earth's mantle and crust is naturally solid. However, there are a few minor pockets of magma that form near the surface where geologic processes cause melting. It is this magma that becomes the source for volcanoes and igneous rocks.

Lava cools quickly on the surface of the earth and forms tiny microscopic crystals. These are known as fine-grained extrusive, or volcanic, igneous rocks. Extrusive rocks are often vesicular, filled with holes from escaping gas bubbles. Volcanism is the process in which lava is erupted. Depending on the properties of the lava that is erupted, the volcanism can be drastically different, from smooth and gentle to dangerous and explosive. This leads to different types of volcanoes and different volcanic hazards.

Lava flow.

In contrast, magma that cools slowly below the earth's surface forms larger crystals which can be seen with the naked eye. These are known as coarse-grained intrusive, or plutonic, igneous rocks. This relationship between cooling rates and grain sizes of the solidified minerals in igneous rocks is important for interpreting the rock's geologic history.

Half Dome, a mass of intrusive igneous rock in Yosemite National Park, now exposed by erosion.

Classification of Igneous Rocks

Igneous rocks are classified based on texture and composition. Texture describes the physical characteristics of the minerals, such as grain size. This relates to the cooling history of the molten magmafrom which it came. Composition refers to the rock's specific mineralogy and chemical composition. Cooling history is also related to changes that can occur to the composition of igneous rocks.

Texture

If magma cools slowly, deep within the crust, the resulting rock is called intrusive or plutonic. The slow cooling process allows crystals to grow large, giving intrusive igneous rock a coarse-grained or phaneritic texture. The individual crystals in phaneritic texture are readily visible to the unaided eye.

Granite is a classic coarse-grained (phaneritic) intrusive igneous rock. The different colors are unique minerals. The black colors are likely two or three different minerals.

Basalt is a classic fine-grained (aphanitic) extrusive igneous rock. This sample is mostly fine groundmass with a few small green phenocrysts that are the mineral olivine.

When lava is extruded onto the surface, or intruded into shallow fissures near the surface and cools, the resulting igneous rock is called extrusive or volcanic. Extrusive igneous rocks have a fine-grained or aphanitic texture, in which the grains are too small to see with the unaided eye. The fine-grained texture indicates the quickly cooling lava did not have time to grow large crystals. These tiny crystals can be viewed under a petrographic microscope. In some cases, extrusive lava cools so rapidly it does not develop crystals at all. This non-crystalline material is not classified as minerals, but as volcanic glass. This is a common component of volcanic ash and rocks like obsidian.

Porphyritic texture.

Some igneous rocks have a mix of coarse-grained minerals surrounded by a matrix of fine-grained material in a texture called porphyritic. The large crystals are called phenocrysts and the fine-grained matrix is called the groundmass or matrix. Porphyritic texture indicates the magma body underwent a multi-stage cooling history, cooling slowly while deep under the surface and later rising to a shallower depth or the surface where it cooled more quickly.

Pegmatitic texture.

Residual molten material expelled from igneous intrusions may form veins or masses containing very large crystals of minerals like feldspar, quartz, beryl, tourmaline, and mica. This texture, which indicates a very slow crystallization, is called pegmatitic. A rock that chiefly consists of pegmatitic texture is known as a pegmatite. To give an example of how large these crystals can get, transparent cleavage sheets of pegmatitic muscovite mica were used as windows during the Middle Ages.

Scoria, a vesicular extrusive igneous rock.

Pumice.

All magmas contain gases dissolved in solution called volatiles. As the magma rises to the surface, the drop in pressure causes the dissolved volatiles to come bubbling out of solution, like the fizz in an opened bottle of soda. The gas bubbles become trapped in the solidifying lava to create a vesicular texture, with the holes specifically calledo vesicles. The type of volcanic rock with common vesicles is called scoria.

Lava that cools extremely quickly may not form crystals at all, even microscopic ones. The resulting rock is called volcanic glass. Obsidian is a rock consisting of volcanic glass. Obsidian as a glassy rock shows an excellent example of conchoidal fracture similar to the mineral quartz.

Obsidian (volcanic glass). Note conchoidal fracture.

When volcanoes erupt explosively, vast amounts of lava, rock, ash, and gases are thrown into the atmosphere. The solid parts, called tephra, settle back to earth and cool into rocks with pyroclastic textures. Pyro, meaning fire, refers to the igneous source of the tephra and clastic refers to the rock fragments. Tephra fragments are named based on size—ash (<2 mm), lapilli (2-64 mm), and bombs or blocks (>64 mm). Pyroclastic texture is usually recognized by the chaotic mix of crystals, angular glass shards, and rock fragments. Rock formed from large deposits of tephra fragments is called tuff. If the fragments accumulate while still hot, the heat may deform the crystals and weld the mass together, forming a welded tuff.

This tuff has crystals, rock fragments, and bedrock mixed together.

Composition

Composition refers to a rock's chemical and mineral make-up. For igneous rock, composition is divided into four groups: felsic, intermediate, mafic, and ultramafic. These groups refer to differing amounts of silica, iron, and magnesium found in the minerals that make up the rocks. It is important to realize these groups do not have sharp boundaries in nature, but rather lie on a continuous spectrum with many transitional compositions and names that refer to specific quantities of minerals. As an example, granite is a commonly-used term, but has a very specific definition which includes exact quantities of minerals like feldspar and quartz. Rocks labeled as 'granite' in laymen applications can be several other rocks, including syenite, tonalite, and monzonite. To avoid these complications, the following figure presents a simplified version of igneous rock nomenclature focusing on the four main groups, which is adequate for an introductory student.

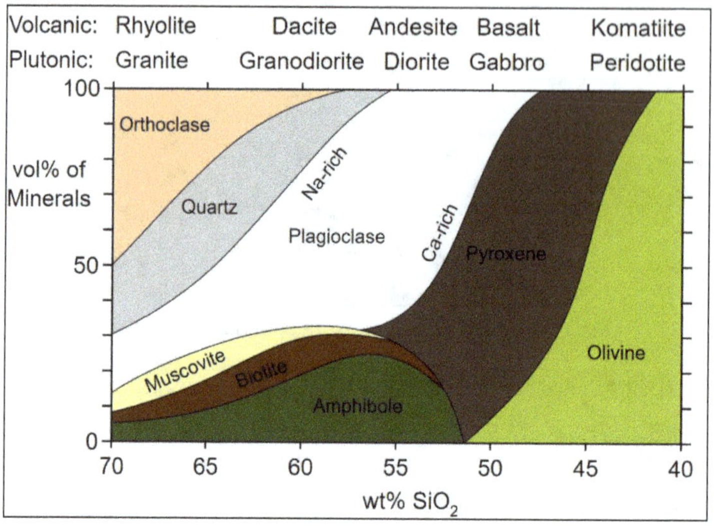

Mineral composition of common igneous rocks. Percentage of minerals is shown on the vertical axis. Percentage of silica is shown on the horizontal axis. Rock names at the top include a continuous spectrum of compositions grading from one into another.

Felsic refers to a predominance of the light-colored (felsic) minerals feldspar and silica in the form of quartz. These light-colored minerals have more silica as a proportion of their overall chemical formula. Minor amounts of dark-colored (mafic) minerals like amphibole and biotite mica may be present as well. Felsic igneous rocks are rich in silica (in the 65-75% range, meaning the rock would be 65-75% weight percent SiO_2) and poor in iron and magnesium.

Intermediate is a composition between felsic and mafic. It usually contains roughly-equal amounts of light and dark minerals, including light grains of plagioclase feldspar and dark minerals like amphibole. It is intermediate in silica in the 55-60% range.

Mafic refers to an abundance of ferromagnesian minerals (with magnesium and iron, chemical symbols Mg and Fe) plus plagioclase feldspar. It is mostly made of dark minerals like pyroxene and olivine, which are rich in iron and magnesium and relatively poor in silica. Mafic rocks are low in silica, in the 45-50% range.

Ultramafic refers to the extremely mafic rocks composed of mostly olivine and some pyroxene which have even more magnesium and iron and even less silica. These rocks are rare on the surface, but make up peridotite, the rock of the upper mantle. It is poor in silica, in the 40% or less range.

In the figure above, the top row has both plutonic and volcanic igneous rocks arranged in a continuous spectrum from felsic on the left to intermediate, mafic, and ultramafic toward the right. Rhyolite thus refers to the volcanic and felsic igneous rocks, and granite thus refer to intrusive and felsic igneous rocks. Andesite and diorite likewise refer to extrusive and intrusive intermediate rocks (with dacite and granodiorite applying to those rocks with composition between felsic and intermediate). Basalt and gabbro are the extrusive and intrusive names for mafic igneous rocks, and peridotite is ultramafic, with komatiite as the fine-grained extrusive equivalent. Komatiite is a rare rock because volcanic material that comes direct from the mantle is not common, although some examples can be found in ancient Archean rocks. Nature rarely has sharp boundaries and the classification and naming of rocks often imposes what appear to be sharp boundary names onto a continuous spectrum.

TEXTURE — Match the texture with the composition of the rock to identify it.	Felsic >10% quartz >50% feldspar <15% mafic minerals	Intermediate >10% quartz >50% plagioclase >10% orthoclase <50% mafic minerals	Mafic 20-85% plagioclase 15-50% pyroxene >35% olivine	Ultramafic Olivine & pyroxene
Phaneritic (coarse grained)	Granite	Diorite	Gabbro	Peridotite
Aphanitic (fine grained)	Rhyolite	Andesite	Basalt	
Porphyritic (large crystals in a fine matrix)	Porphyritic Rhyolite	Porphyritic Andesite	Porphyritic Basalt	
Vesicular (bubbly or frothy)	Pumice		Scoria	
Glassy	Obsidian			
Pyroclastic (fragmental)	Rhyolitic Tuff or Volcanic Breccia	Andesitic Tuff or Volcanic Breccia	Basaltic Tuff or Volcanic Breccia	

Igneous rock classification table with composition as vertical columns and texture as horizontal rows.

Aphanitic/Phaneritic Rock Types

Felsic Composition

Granite from Cape Cod.

- Granite is a course-crystalline felsic intrusive rock. The presence of quartz is a good indicator of granite. Granite commonly has large amounts of salmon pink potassium feldspar and white plagioclase crystals that have visible cleavage planes. Granite is a good approximation for the continental crust, both in density and composition.

Rhyolite.

- Rhyolite is a fine-crystalline felsic extrusive rock. Rhyolite is commonly pink and will often have glassy quartz phenocrysts. Because felsic lavas are less mobile, it is less common than granite. Examples of rhyolite include several lava flows in Yellowstone National Park and the altered rhyolite that makes up the Grand Canyon of the Yellowstone.

Intermediate Composition

Diorite.

- Diorite is a coarse-crystalline intermediate intrusive igneous rock. Diorite is identifiable by it's Dalmatian-like appearance of black hornblende and biotite and white plagioclase feldspar. It is found in its namesake, the Andes Mountains as well as the Henry and Abajo mountains of Utah.
- Andesite is a fine crystalline intermediateextrusive rock. It is commonly grey and porphyritic. It can be found in the Andes Mountains and in some island arcs. It is the fine grained compositional equivalent of diorite.

Andesite.

Mafic Composition

Gabbro.

- Gabbro is a coarse-grained mafic igneous rock, made with mainly mafic minerals like pyroxene and only minor plagioclase. Because mafic lava is more mobile, it is less common than basalt. Gabbro is a major component of the lower oceanic crust.

Vesicular Basalt.

- Basalt is a fine-grained mafic igneous rock. It is commonly vesicular and aphanitic. When porphyritic, it often has either olivine or plagioclase phenocrysts.

Basalt is the main rock which is formed at mid-ocean ridges, and is therefore the most common rock on the Earth's surface, making up the entirety of the ocean floor (except where covered by sediment).

Igneous Rock Bodies

Igneous rocks are common in the geologic record, but surprisingly, it is the intrusive rocks that are more common. Extrusive rocks, because of their small crystals and glass, are less durable. Plus, they are, by definition, exposed to the elements of erosion immediately. Intrusive rocks, forming underground with larger, stronger crystals, are more likely to last. Therefore, most landforms and rock groups that owe their origin to igneous rocks are intrusive bodies. A significant exception to this is active volcanoes.

Dike of olivine gabbro cuts across Baffin Island.

When magma intrudes into a weakness like a crack or fissure and solidifies, the resulting cross-cutting feature is called a dike (sometimes spelled dyke). Because of this, dikes are often vertical or at an angle relative to the pre-existing rock layers that they intersect. Dikes are therefore discordant intrusions, not following any layering that was present. Dikes are important to geologists, not only for the study of igneous rocks themselves but also for dating rock sequences and interpreting the geologic history of an area.

Igneous sill intruding between Paleozoic strata in Nova Scotia.

A magma chamber is a large underground reservoir of molten rock. The path of rising magma is called a diapir. The processes by which a diapir intrudes into the surrounding native or country rock are not well understood and are the subject of ongoing geological inquiry. For example, it is not known what happens to the pre-existing country rock as the diapir intrudes. One theory is the overriding rock gets shouldered aside, displaced by the increased volume of magma. Another is the native rock is melted and consumed into the rising magma or broken into pieces that settle into the magma, a process known as stoping. It has also been proposed that diapirs are not a real phenomenon, but just a series of dikes that blend into each other. The dikes may be intruding over millions of years, but since they may be made of similar material, they would be appearing to be formed at the same time. Regardless, when a diapir cools, it forms a mass of intrusive rock called a pluton. Plutons can have irregular shapes, but can often be somewhat round.

Cottonwood Stock, a quartz monzonite pluton exposed at the mouth of Little Cottonwood Canyon.

Half Dome in Yosemite National Park is a part of the Sierra Nevada batholith which is mostly made of granite.

When many plutons merge together in an extensive single feature, it is called a batholith. Batholiths are found in the cores of many mountain ranges, including the granite formations of Yosemite National Park in the Sierra Nevada of California. They are typically more than 100 km2 in area, associated with subduction zones, and mostly felsic in composition. A stock is a type of pluton with less surface exposure than a batholith, and may represent a narrower neck of material emerging from the top of a batholith. Batholiths and stocks are discordant intrusions that cut across and through surrounding country rock.

The Henry Mountains in Utah are interpreted to be a laccolith, exposed by erosion of the overlying layers.

Laccoliths are blister-like, concordant intrusions of magma that form between sedimentary layers. The Henry Mountains of Utah are a famous topographic landform formed by this process. Laccoliths bulge upwards; a similar downward-bulging intrusion is called a lopolith.

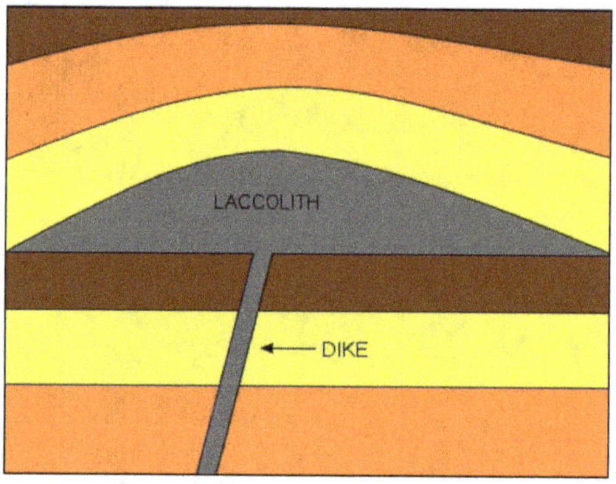

Laccolith forms as a blister in between sedimentary strata.

Aeolian Process

Wind erosion of soil at the foot of Chimborazo.

Rock carved by drifting sand below Fortification Rock.

Aeolian processes, also spelled eolian or æolian, pertain to wind activity in the study of geology and weather and specifically to the wind's ability to shape the surface of the Earth (or other planets). Winds may erode, transport, and deposit materials and are effective agents in regions with sparse vegetation, a lack of soil moisture and a large supply of unconsolidated sediments. Although water is a much more powerful eroding force than wind, aeolian processes are important in arid environments such as deserts.

Wind Erosion

Wind erodes the Earth's surface by deflation (the removal of loose, fine-grained particles by the turbulent action of the wind) and by abrasion (the wearing down of surfaces by the grinding action and sandblasting by windborne particles).

Regions which experience intense and sustained erosion are called deflation zones. Most aeolian deflation zones are composed of desert pavement, a sheet-like surface of rock fragments that remains after wind and water have removed the fine particles. Almost half of Earth's desert surfaces are stony deflation zones. The rock mantle in desert pavements protects the underlying material from deflation.

A dark, shiny stain, called desert varnish or rock varnish, is often found on the surfaces of some desert rocks that have been exposed at the surface for a long period of time. Manganese, iron oxides, hydroxides, and clay minerals form most varnishes and provide the shine.

A rock sculpted by wind erosion in the Altiplano region of Bolivia.

Sand blowing off a crest in the Kelso Dunes of the Mojave Desert, California.

Wind-carved alcove in the Navajo Sandstone near Moab, Utah.

Deflation basins, called blowouts, are hollows formed by the removal of particles by wind. Blowouts are generally small, but may be up to several kilometers in diameter.

Wind-driven grains abrade landforms. In parts of Antarctica wind-blown snowflakes that are technically sediments have also caused abrasion of exposed rocks. Grinding by particles carried in the wind creates grooves or small depressions. Ventifacts are rocks which have been cut, and sometimes polished, by the abrasive action of wind.

Sculpted landforms, called yardangs, are up to tens of meters high and kilometers long and are forms that have been streamlined by desert winds. The famous Great Sphinx of Giza in Egypt may be a modified yardang.

List of Major Aeolian Movements

Major global aeolian dust movements thought to influence and be influenced by weather and climate variation:

- From Sahara (specifically Sahel and Bodélé Depression) averaged 182 million tons of dust each year between 2007 and 2011, carried past the western edge of the Sahara at longitude 15W. Variation: 86%. Destination: 132 mln tons cross the Atlantic (avg), 27.7 mln tons fall in Amazon Basin (avg), 43 mln make it to the Caribbean. Texas and Florida also receive the dust. Events have become far more common in recent decades. NASA's Cloud-Aerosol Lidar and Infrared Pathfinder Satellite Observation (CALIPSO) data. Harmattan winter dust storms in West Africa also occur blowing dust to the ocean.
- Gobi Desert to Korea, Japan, Taiwan (at times) and even Western USA (blowing east).
- Thar Desert pre-monsoon towards Delhi, Uttar Pradesh, Indo-Gangetic Plain.
- Shamal June–July winds blowing dust in primarily north to south in Saudi Arabia, Iran, Iraq, UAE, and parts of Pakistan.
- Haboob dust storms in Sudan, Australia, Arizona associated with monsoon.

- Khamsin dust from Libya, Egypt and Levant in Spring associated with extra-tropical cyclones.
- Dust Bowl event in USA, carried sand eastward. 5500 tons were deposited in Chicago area.
- Sirocco sandy winds from Africa/Sahara blowing north into South Europe.
- Kalahari Desert blowing east to southern Indian Ocean and Australia.

Transport

Dust storm approaching Spearman, Texas.

A massive sand storm cloud is about to envelop a military camp as it rolls over Al Asad, Iraq, just before nightfall.

Particles are transported by winds through suspension, saltation (skipping or bouncing) and creeping (rolling or sliding) along the ground.

Small particles may be held in the atmosphere in suspension. Upward currents of air support the weight of suspended particles and hold them indefinitely in the surrounding air. Typical winds near Earth's surface suspend particles less than 0.2 millimeters in diameter and scatter them aloft as dust or haze.

Dust storm in Amarillo, Texas.

Saltation is downwind movement of particles in a series of jumps or skips. Saltation normally lifts sand-size particles no more than one centimeter above the ground and proceeds at one-half to one-third the speed of the wind. A saltating grain may hit other grains that jump up to continue the saltation. The grain may also hit larger grains that are too heavy to hop, but that slowly creep forward as they are pushed by saltating grains. Surface creep accounts for as much as 25 percent of grain movement in a desert.

Aeolian turbidity currents are better known as dust storms. Air over deserts is cooled significantly when rain passes through it. This cooler and denser air sinks toward the desert surface. When it reaches the ground, the air is deflected forward and sweeps up surface debris in its turbulence as a dust storm.

Crops, people, villages, and possibly even climates are affected by dust storms. Some dust storms are intercontinental, a few may circle the globe, and occasionally they may engulf entire planets. When the Mariner 9 spacecraft entered its orbit around Mars in 1971, a dust storm lasting one month covered the entire planet, thus delaying the task of photo-mapping the planet's surface.

Most of the dust carried by dust storms is in the form of silt-size particles. Deposits of this windblown silt are known as loess. The thickest known deposit of loess, 335 meters, is on the Loess Plateau in China. This very same Asian dust is blown for thousands of miles, forming deep beds in places as far away as Hawaii. In Europe and in the Americas, accumulations of loess are generally from 20 to 30 meters thick. The soils developed on loess are generally highly productive for agriculture.

Aeolian transport from deserts plays an important role in ecosystems globally, e.g. by transport of minerals from the Sahara to the Amazon basin. Saharan dust is also responsible for forming red clay soils in southern Europe. Aeolian processes are affected by human activity, such as the use of 4x4 vehicles.

Small whirlwinds, called dust devils, are common in arid lands and are thought to be related to very intense local heating of the air that results in instabilities of the air mass. Dust devils may be as much as one kilometer high.

Deposition

Wind-deposited materials hold clues to past as well as to present wind directions and intensities. These features help us understand the present climate and the forces that molded it. Wind-deposited sand bodies occur as sand sheets, ripples, and dunes.

Sand sheets are flat, gently undulating sandy plots of sand surfaced by grains that may be too large for saltation. They form approximately 40 percent of aeolian depositional surfaces. The Selima Sand Sheet in the eastern Sahara Desert, which occupies 60,000 square kilometers in southern Egypt and northern Sudan, is one of the Earth's largest sand sheets. The Selima is absolutely flat in a few places; in others, active dunes move over its surface.

Wind blowing on a sand surface ripples the surface into crests and troughs whose long axes are perpendicular to the wind direction. The average length of jumps during saltation corresponds to the wavelength, or distance between adjacent crests, of the ripples. In ripples, the coarsest materials collect at the crests causing inverse grading. This distinguishes small ripples from dunes, where the coarsest materials are generally in the troughs. This is also a distinguishing feature between water laid ripples and aeolian ripples.

Accumulations of sediment blown by the wind into a mound or ridge, dunes have gentle upwind slopes on the windward side. The downwind portion of the dune, the lee slope, is commonly a steep avalanche slope referred to as a slipface. Dunes may have more than one slipface. The minimum height of a slipface is about 30 centimeters.

Wind-blown sand moves up the gentle upwind side of the dune by saltation or creep. Sand accumulates at the brink, the top of the slipface. When the buildup of sand at the brink exceeds the angle of repose, a small avalanche of grains slides down the slipface. Grain by grain, the dune moves downwind.

Some of the most significant experimental measurements on aeolian sand movement were performed by Ralph Alger Bagnold, a British army engineer who worked in Egypt prior to World War II. Bagnold investigated the physics of particles moving through the atmosphere and deposited by wind. He recognized two basic dune types, the crescentic dune, which he called "barchan", and the linear dune, which he called longitudinal or "seif".

A study published in Catena examined the effect of vegetation on aeolian dust accumulation in the semiarid steppe of northern China. Using a series of trays with different vegetation coverage and a control model with none, the authors found that an increase in vegetation coverage improves the efficiency of dust accumulation and adds more nutrients to the environment, particularly organic carbon. Two critical point were revealed by their data:

- The efficiency of trapping dust increases slowly above 15% coverage, and decreases rapidly below 15% coverage.
- At around 55%-75% coverage, dust accumulation reaches a maximum capacity.

In Europe, the European Commission requested the Joint Research Centre to develop the first pan-European wind erosion map. In a first step, a group of scientists have used the LUCAS topsoil dataset to develop the wind erosion susceptibility of European soils. Then, they have developed an index to land susceptibility for making a qualitative assessment of wind erosion. Finally, they modified the RWEQ model to estimate the soil loss due to wind erosion in European Agricultural soils.

A three-year quantitative study on the effects of vegetation removal on wind erosion found that the removal of grasses in an aeolian environment increased the rate of soil deposition. In the same study, a relationship was shown between decreasing plant density with decreasing soil nutrients. Similarly, horizontal soil flux across the test site was shown to increase with increasing vegetation removal.

Cross-bedding of sandstone near Mount Carmel road, Zion National Park, indicating wind action and sand dune formation prior to formation of rock.

A study published in Earth Surfaces Processes and Landforms investigated the relationship between vegetative cover on sand surfaces with the rate of sand transport. It was found that sand flux decreased exponentially with vegetation cover. This was done by measuring plots of land with varying degrees of vegetation against rates of sand transport. The authors contend that this relationship can be utilized to manipulate rates of sediment flux by introducing vegetation in an area or to quantify human impact by recognizing vegetation loss's effect on sandy landscapes.

Mesquite Flat Dunes in Death Valley looking toward the Cottonwood Mountains from the north west arm of Star Dune.

Holocene eolianite deposit on Long Island, The Bahamas. This unit is formed of wind-blown carbonate grains.

Aeolian deposition near Addeha, Kola Tembien.

Erosion and Tectonics

The interaction between erosion and tectonics has been a topic of debate since the early 1990s. While the tectonic effects on surface processes such as erosion have long been recognized (for example, river formation as a result of tectonic uplift), the opposite (erosional effects on tectonic activity) has only recently been addressed.

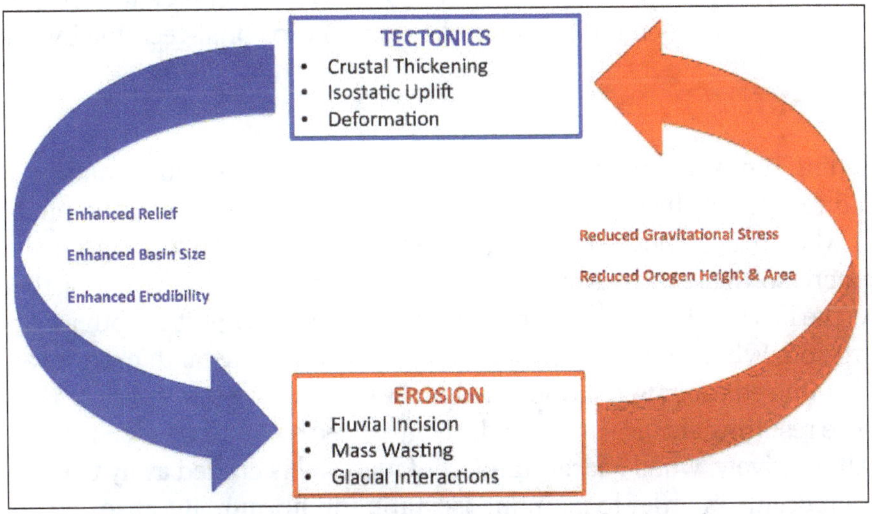

Interactions and feedback pathways for tectonics and erosional processes.

The primary questions surrounding this topic are what types of interactions exist between erosion and tectonics and what are the implications of these interactions. While this is still a matter of debate, one thing is clear, the Earth's landscape is a product of two factors: tectonics, which can create topography and maintain relief through surface and rock uplift, and climate, which mediates the erosional processes that wear away upland areas over time. The interaction of these processes can form, modify, or destroy geomorphic features on the Earth's surface.

Tectonic Processes

The term tectonics refers to the study of the Earth's surface structure and the ways in which it changes over time. Tectonic processes typically occur at plate boundaries which are one of three types: convergent boundaries, divergent boundaries, or transform boundaries. These processes form and modify the topography of the Earth's surface, effectively increasing relief through the mechanisms of isostatic uplift, crustal thickening, and deformation in the form of faulting and folding. Increased elevations, in relation to regional base levels, lead to steeper river channel gradients and an increase in orographically localized precipitation, ultimately resulting in drastically increased erosion rates. The topography, and general relief, of a given area determines the velocity at which surface runoff will flow, ultimately determining the potential erosive power of the runoff. Longer, steeper slopes are more prone to higher rates of erosion during periods of heavy rainfall than shorter, gradually sloping areas. Thus, large mountain ranges, and other areas of high relief, formed through tectonic uplift will have significantly higher rates of erosion. Additionally, tectonics can directly influence erosion rates on a short timescale, as is clear in the case of earthquakes, which can trigger landslides and weaken surrounding rock through seismic disturbances.

While tectonic uplift in any case will lead to some form of increased elevation, thus higher rates of erosion, a primary focus is set on isostatic uplift as it provides a fundamental connection between the causes and effects of erosional-tectonic interactions.

Isostatic Uplift

Understanding the principle of isostasy is a key element to understanding the interactions and feedbacks shared between erosion and tectonics. The principle of isostasy states that when free to move vertically, lithosphere floats at an appropriate level in the asthenosphere so that the pressure at a depth of compensation in the asthenosphere well below the base of the lithosphere is the same. Isostatic uplift is both a cause and an effect of erosion. When deformation occurs in the form of crustal thickening an isostatic response is induced causing the thickened crust to sink, and surrounding thinner crust to uplift. The resulting surface uplift leads to enhanced elevations, which in turn induces erosion. Alternatively, when a large amount of material is eroded away from the Earth's surface uplift occurs in order to maintain isostatic equilibrium. Because of isostasy, high erosion rates over significant horizontal areas can effectively suck up material from the lower crust and upper mantle. This process is known as isostatic rebound and is analogous to Earth's response following the removal of large glacial ice sheets.

Isostatic uplift and corresponding erosion are responsible for the formation of regional-scale geologic features as well as localized structures. Two such examples include:

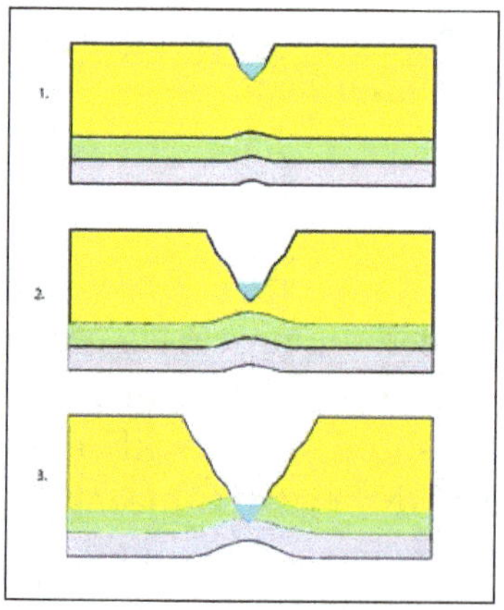

Formation of a river anticline.

- Continental shields: Generally large areas of low relief (<100 m) in the Earth's crust where Precambrian crystalline igneous and high-grade metamorphic rocks are exposed. Shields are considered tectonically stable areas in comparison to the activity occurring at their margins and the boundaries between plates, but their formation required large amounts of tectonic activity and erosion. Shields, along with stable platforms, are the basic tectonic components of continents, therefore understanding their development is critical to understanding the development of other surface features on Earth. Initially, a mountain belt is formed at a convergent plate margin. Transformation of a mountain belt to a shield is majorly dependent on two factors: (1) erosion of the mountain belt by running water and (2) isostatic adjustment resulting from the removal of surface rock due to erosion. This process of erosion followed by isostatic adjustment continues until the system is at isostatic equilibrium. At this point large-scale erosion can no longer occur because the surface has eroded down to nearly sea-level and uplift ceases due to the system's state of equilibrium.

- River anticlines: Geologic structures formed through the focused uplift of rock underlying confined areas of high erosion (i.e., rivers). Isostatic rebound resulting from the rapid removal of overlying rock, via erosion, causes the weakened areas of crustal rock to uplift from the apex of the river. In order for the development of these structures to occur the erosion rate of the river must exceed both the average erosional rate of the area, and the rate of uplift of the orogen. The two factors influencing the development of these structures are stream power of the associated river and the flexural rigidity of the crust in the area. The combination of increased stream power with decreased flexural rigidity results in the system's progression from a transverse anticline to a river anticline.

Channel Flow

Channel flow describes the process through which hot, viscous crustal material flows horizontally between the upper crust and lithospheric mantle, and is eventually pushed to the surface. This model aims to explain features common to metamorphic hinterlands of some collisional orogens, most notably the Himalaya–Tibetan plateau system. In mountainous areas with heavy rainfall (thus, high erosion rates) deeply incising rivers will form. As these rivers wear away the Earth's surface two things occur: (1) Pressure is reduced on the underlying rocks effectively making them weaker and (2) The underlying material moves closer to the surface. This reduction of crustal strength, coupled with the erosional exhumation, allows for the diversion of the underlying channel flow toward the Earth's surface.

Erosional Processes

The term erosion refers to the group of natural processes, including weathering, dissolution, abrasion, corrosion, and transportation, by which material is worn away from the Earth's surface to be transported and deposited in other locations.

A natural arch produced by erosion of differentially weathered rock in Jebel Kharaz.

- Differential erosion: Erosion that occurs at irregular or varying rates, caused by the differences in the resistance and hardness of surface materials; softer and weaker rocks are rapidly worn away, whereas harder and more resistant rocks remain to form ridges, hills, or mountains. Differential erosion, along with the tectonic setting, are two of the most important controls on the evolution of continental landscapes on Earth.

The feedback of erosion on tectonics is given by the transportation of surface, or near-surface, mass (rock, soil, sand, regolith, etc.) to a new location. This redistribution of material can have profound effects on the state of gravitational stresses in the area, dependent on the magnitude of mass transported. Because tectonic processes are highly dependent on the current state of gravitational stresses, redistribution of surface material can lead to tectonic activity. While erosion in all of its forms, by definition, wears away material from the Earth's surface, the process of mass wasting as a product of deep fluvial incision has the highest tectonic implications.

Mass Wasting

Talus cones produced by mass wasting, north shore of Isfjord, Svalbard, Norway.

Mass wasting is the geomorphic process by which surface material move downslope typically as a mass, largely under the force of gravity. As rivers flow down steeply sloping mountains, deep channel incision occurs as the river's flow wears away the underlying rock. Large channel incision progressively decreases the amount of gravitational force needed for a slope failure event to occur, eventually resulting in mass wasting. Removal of large amounts of surface mass in this fashion will induce an isostatic response resulting in uplift until equilibrium is reached.

Effects on Structural Evolution

Recent studies have shown that erosional and tectonic processes have an effect on the structural evolution of some geologic features, most notably orogenic wedges. Highly useful sand box models, in which horizontal layers of sand are slowly pressed against a backstop, have shown that the geometries, structures, and kinematics of orogenic wedge formation with and without erosion and sedimentation are significantly different. Numerical models also show that the evolution of orogens, their final tectonic structure, and the potential development of a high plateau, all are sensitive to the long term climate over the mountains, for example, the concentration of precipitation in one side of the orogen due to orographic lift under a dominant wind direction.

Cycle of Erosion

The geographic cycle or cycle of erosion is an idealized model that explains the development of relief in landscapes. The model starts with the erosion that follows uplift of land above a base level and ends – if conditions allow – in the formation of a peneplain. Landscapes that show evidence of more than one cycle of erosion are termed "polycyclical". The cycle of erosion and some of its associated concepts have, despite popularity, been a subject of much criticism.

There had been some ideas on cyclical erosion in the Graeco-Roman world and then in the Islamic world and Europe during the Middle Ages. However the immediate influences of William Morris Davis, the creator of the cycle of erosion model, were 19th century American explorers. The end of the American Civil War led to a resumption of the exploration of the western United States. Three explorers, John Wesley Powell, Clarence Edward Dutton and Grove Karl Gilbert, wrote about the geomorphology and geology in the landscapes they encountered. It was from these works that Davis borrowed many concepts used to construct the model.

It has been argued that Davis was also influenced by ideas from the field of biology, especially the Neo-Lamarckian thought that was current in the late 19th-century United States. It is thought that Davis received some of this Neo-Lamarckian influence from his tutor, Nathaniel Shaler. Other biological theories that may have shaped the cycle of erosion are those of orthogenesis and recapitulation both of which are linked to Neo-Lamarckianism. Darwin's evolution theory was a lesser influence relative to Neo-Lamarckism. The impact of these ideas can be hinted in the models' employment of the concept of "evolution" rather than "change", implying a predictable direction of landscape and landform change.

Davis conceived the cycle of erosion model in the 1880s. More fully developed by 1900 it received wide acclaim, but was never universally accepted. The initial enthusiasm and strength of the cycle of erosion model has been attributed to various causes. First, the model provided a framework to study areas and epochs in Earth history where erosion is the dominant process. Second, the model fitted well into the grand evolutionary thought that had emerged in the 19th century with Darwin's evolution theory. Lastly, some popularity can be attributed to Davis's lucid writing style. The model achieved its greatest popularity in the 1900–39 period when numerous studies on denudation chronology based on the model were published. In these studies usually two to five erosion cycles were identified. The approach of doing denudation chronology with the cycle of erosion model lost popularity from the 1930s onward. Arguably this was so because the approach did not provide any unforeseeable insights. An increasing number of geomorphologists had begun to study processes happening in the present and not in the past as done with the cycle of erosion model. These process geomorphologists soon realized some of their observations were at odds with Davis's model. Other geomorphologists turned away from the cycle of erosion to work instead on climatic or tectonic geomorphology.

The model spread fast. In 1901 Hans Reusch was using it to explain the undulating plateau of southern Norway. Very much influenced by Davis Walter Wråk moved to study the relief of the northern Scandinavian Mountains, describing among other things the Borsu surface. The first study of China's topography using the model was published in 1907 by B. Willis and co-workers.

While the model was at first widely accepted among Anglo-Saxon scholars, in Continental Europe it met some resistance with German scholars Albrecht and Walther Penck, Siegfried Passarge and Alfred Hettner standing out as early opponents to the model. Despite Davis's efforts, which included translating his work into German, being a visiting professor at Berlin and touring much of Europe, the cycle of erosion never did take firm root in Germany. Walther Penck came to originate an alternative model with its own set of tenets. Walther's ideas were published posthumously in 1924 by his father, Albrecht. In Walther's model, relief is shaped as a function of the denudation/uplift rate. Davis, a friend of Walther's father, was critical of Walther's new model. Davis's review of Walther Penck's second publication on the subject left a distorted view of Penck's ideas among later workers. This is because Davis misunderstood and mis-translated parts of the paper, therefore Davis is not an adequate critic of Penck's work.

An alternative cycle theory was later proposed by Lester Charles King.

Intensified Criticism

While King's ideas were an attempt at refuting Davis's cycle of erosion they were themselves of cyclical nature and contributed to ushering in a wave of criticism in against both his and Davis's models. This criticism has been called "Davis bashing" by Cliff Ollier and constitutes to Ollier the ridicule of cyclical theories in geomorphology without any alternative model being proposed. Since Davis' ideas were being discredited other areas of research, like that of climatic geomorphology, were attacked by their perceived association to it.

Richard Chorley, a geomorphologist known for his criticism of the cycle of erosion.

The notions of time, uplift, slope and drainage density evolution in the erosion cycle have been criticized. Further the validity of some whole concepts associated with the cycle of erosion have been questioned including stream grade, base level and most of all that of peneplains.

Currently geomorphologists are divided into three camps: those who still adhere to Davisian concepts, either in the original or a modified form; those who desire to replace Davisian ideas by a different cyclic erosional model; and those who reject cyclic erosion completely.

Modern Status

Despite considerable criticism the cycle of erosion model has remained part of the science of geomorphology. The model or theory has never been proved wrong, but neither has it been proven. The inherent difficulties of the model have instead made geomorphological research to advance along other lines. In contrast to its disputed status in geomorphology, the cycle of erosion model is a common approach used to establish denudation chronologies, and is thus an important concept in the science of historical geology. While acknowledging its shortcomings modern geomorphologists Andrew Goudie and Karna Lidmar-Bergström have praised it for its elegance and pedagogical value respectively:

> "Stripped of its evolutionary baggage, the Davisian Cycle of Erosion had merit as an interpretive exercise, and indeed still has merit as an end-member in a range of possible temporal scenarios for Earth's surface development."

The Model

William Morris Davis, the originator of the model, divided it into stages whose transition is gradual. The model begins with an uplifted or to-be-uplifted landscape. Then Davis defined a youthful stage where river incision is the dominant process shaping the landscape. During the youthful stage height differences between uplands and valley bottoms increase rapidly. The youthful stage is followed by a mature stage where height differences between valley bottoms and uplands are at their greatest. Beginning in the mature stage slope decline becomes a more important phenomenon, and uplands lose height more rapidly than rivers incise, effectively diminishing relief. In the very latest stage erosion has acted so long that the landscape – despite original height – is reduced into a rolling lowland. This landscape of low relief is called a peneplain and may contain residual heights standing out from the general level. The peneplain can be uplifted, starting a second erosion cycle.

Davis acknowledged that a full cycle was a special case and that initial uplift was not necessarily rapid nor followed by a prolonged period of quiescence. However, as Walther Penck pointed out, Davis and his followers usually –if not almost always– employed a rapid uplift and quiescence approach to explain landscapes. This means that the model, as understood by most, assumes rapid and episodic tectonic uplift. Another characteristic of the model is that slopes evolve by decline, meaning that initially steep slopes are worn out by erosion forming successively gentler slopes. Weaknesses of the model are that it is mostly theoretical and deductive in nature; further it does not take into account the complexity of tectonic movements or climate change. The nature of surface processes is also poorly represented by the model. The model in its original form is intended to explain relief development in temperate landscapes where erosion by running water is assumed to be of prime importance. Nevertheless, the cycle of erosion has been extended, with modifications, into arid, glacial, coastal, karst and periglacial areas.

Variants of the Cycle of Erosion

Arid

At the beginning of the cycle of erosion in arid climate there are numerous small basins to where material is washed during the scarce rainfall events. In the next stage (youthful stage) valleys are developed and highlands dissected by these. Gentle slopes and basins accumulated material derived from the highlands. In the mature stage drainage basins coalesce. At the end a stage is reached where the terrain has lost much of its relief and deflation hollows interfere with the drainage systems, breaking it up into local systems. During all stages sand and dust might be exported by wind to other landscapes.

Coastal

Alternate models are proposed for shore profiles: one for shore profiles of emergence and another for shore profiles of submergence. A complementary model can be applied to shore lines where different parts might have undergone submergence and emergence.

Glacial

The glacial cycle of erosion deals with mountainous regions and lacks an old stage since Davis argues that nothing more developed than mature glacial landscapes exists at present. A glacial cycle of erosion begins with a pre-glacial landscape. Over time valley glaciers erode the underlying rock at different rates, creating valleys and glaciers that are more entrenched than others. As time passes the less-entrenched glaciers reach the same levels of entrenchment as the more entrenched ones, since the deeper a glacier erodes, its erosive power diminishes. In a mature stage, valleys form smooth-sided troughs.

Karst

The cycle of erosion in karst regions has three phases. At first the upper parts of fractures are dissolved, enlarged and filled with water. Normal fluvial valleys develop on the surface, small poljes might exist. Subsequently, re-routing of water by the growth of a karst system disorganizes the fluvial drainage pattern, with valley bottoms developing large dolines and uvalas. Ridges between uvalas gradually disappear as these landforms coalesce. If the bedrock is underlain by insoluble rock, normal fluvial valleys will slowly re-appear once the underground river systems reach the insoluble rock. In the end soluble rocks appear only as isolated hills. Contrary to Davis' original cycle of erosion the karst one does not end in the formation of a peneplain.

Periglacial

The periglacial cycle of erosion begins with a non-periglaciated landscape. Once-periglaciated mass wasting of regolith exposes bedrock in the upper slopes. These outcrops are then subject to frost weathering that makes slopes retreat forming extensive blockfields at the base of the bedrock areas. At a later stage solifluction wears down summits and fills in topographic lows.

Fluvial Processes

Deep, eroding glaciofluvial deposits alongside the Matanuska River, Alaska.

In geography and geology, fluvial processes are associated with rivers and streams and the deposits and landforms created by them. When the stream or rivers are associated with glaciers, ice sheets, or ice caps, the term glaciofluvial or fluvioglacial is used.

Fluvial processes include the motion of sediment and erosion or deposition on the river bed.

Erosion by moving water can happen in two ways. Firstly, the movement of water across the stream bed exerts a shear stress directly onto the bed. If the cohesive strength of the substrate is lower than the shear exerted, or the bed is composed of loose sediment which can be mobilized by such stresses, then the bed will be lowered purely by clear-water flow. However, if the river carries significant quantities of sediment, this material can act as tools to enhance wear of the bed (abrasion). At the same time the fragments themselves are ground down, becoming smaller and more rounded (attrition).

Sediment in rivers is transported as either bedload (the coarser fragments which move close to the bed) or suspended load (finer fragments carried in the water). There is also a component carried as dissolved material.

For each grain size there is a specific velocity at which the grains start to move, called entrainment velocity. However the grains will continue to be transported even if the velocity falls below the entrainment velocity due to the reduced (or removed) friction between the grains and the river bed. Eventually the velocity will fall low enough for the grains to be deposited. This is shown by the Hjulström curve.

A river is continually picking up and dropping solid particles of rock and soil from its bed throughout its length. Where the river flow is fast, more particles are picked up than dropped. Where the river flow is slow, more particles are dropped than picked up. Areas where more particles are dropped are called alluvial or flood plains, and the dropped particles are called alluvium.

Even small streams make alluvial deposits, but it is in the flood plains and deltas of large rivers that large, geologically-significant alluvial deposits are found.

The amount of matter carried by a large river is enormous. The names of many rivers derive from the color that the transported matter gives the water. For example, the Huang He in China is literally translated "Yellow River", and the Mississippi River in the United States is also called "the Big Muddy". It has been estimated that the Mississippi River annually carries 406 million tons of sediment to the sea, the Yellow River 796 million tons, and the Po River in Italy 67 million tons.

Hillslope Processes and Mass Movement

Hillslopes are an important part of the terrestrial landscape. The Earth's landscape can be thought of as being composed of a mosaic of slope types, ranging from steep mountains and cliffs to almost flat plains. On most hillslopes large quantities of soil and sediment are moved over time via the mediums of air, water, and ice often under the direct influence of gravity. The form a hillslope takes is dependent on the various geomorphic processes acting on it. Hillslopes are also the source of materials that are used to construct a number of depositional landforms.

In practical terms, hillslopes have direct and indirect influence on a number of human activities. The steepness and structural stability of hillslopes determines their suitability for agriculture, forestry, and human settlement. Hillslopes can also become a hazard to humans if their materials move rapidly through the process of mass wasting.

Inputs and Outputs to the Hillslope System

We can begin our study of hillslopes by thinking of them as a process-response system. The hillslope system receives inputs of solar radiation, precipitation, solid and dissolved substances from the atmosphere, and unconsolidated sediment derived from the weathering of bedrock. The inputs of unconsolidated sediment are controlled by weathering rates. In general, the warmer the climate the higher the rates of bedrock weathering. Rates of weathering are also influenced by the presence of moisture.

Outputs to hillslopes occur by evapotranspiration, by percolation of water and the movement of dissolved substances into the bedrock, and by removal of sediment by streams by glaciers or by ocean waves and currents. Outputs of debris or sediment from hillslope systems are controlled primarily by the availability of erosional mechanisms to transport material that accumulates at the slope's surface and base. For example, the presence of a stream at the base of a hillslope encourages removal of sediment that moves downslope. If the stream's discharge is too small to handle the debris input, sediment will accumulate at the base of the slope.

The magnitude of hillslope inputs and outputs depends upon a number of factors, including bedrock geology, climate, and the nature of the slope to the broader landscape. The balance between inputs and outputs from the hillslope system exerts a major control over the form of the developing slope. In situations where inputs are the controlling factor, the slope is said to be weathering limited because outputs quickly remove any accumulating debris. Where the potential for weathering is high but outputs are restrained the hillslope system is classified as being transport limited. Landscapes that are transport limited are easily recognized by the presence of a deep soil profile.

Mass Movement and Hillslope Stability

A variety of processes exist by which materials can be moved through the hillslope system. These processes are generically known as mass movement or mass wasting. The operation of mass movement processes relies upon the development of instability in the hillslope system. Under these conditions, failure of the slope material can occur on a range of time scales. Some types of mass movement involve rather rapid, spontaneous events. Sudden failures tend to occur when the stresses exerted on the slope materials greatly exceed their strength for short periods of time. In many cases, type of mass movement is produced by the operation of short term trigger mechanisms. Mass movement can also be a less continuous process that occurs over long periods of time. Slow failures often occur when the applied stresses only just exceed the internal strength of the hillslope system.

What are the sources of the stresses and strength acting within hillslope materials? As we have noted, a major source of stress is the gravitational force. The magnitude of this force is related to the angle of the slope and the weight of hillslope sediments and rock. The following equation models this relationship:

$$F = W \sin Ø$$

where F is gravitational force, W is the weight of the material occurring at some point on the slope, and Ø is the angle of the slope.

The internal strength of the hillslope system varies according to the nature of the materials making up the slope. Hillslopes composed of loose materials, like sand and gravel, derived their internal from frictional resistance which depends on the size, shape, and arrangement of the particles. Hillslopes consisting of silt and clay particles obtain their internal strength from particle cohesion which is controlled by the availability of moisture in the soil. Too much moisture breaks the cohesive bonds and can turn a solid hillslope into a river of mud. Rock slopes generally have the greatest internal strength. Internal strength in these systems is derived from the effects of the solidification and crystallization of magma or the lithification of once loose particles.

The stability of a hillslope depends on the relationship between the stresses applied to the materials that make up the slope and their internal strength. Mass movement occurs when the stresses exceed the internal strength. This condition is not always caused by an increase in stress. In some cases, the internal strength of the materials can be reduced over relatively short periods of time resulting in mass movement.

Many factors can act as triggers for hillslope failure. One of the most common is prolonged or heavy rainfall. Rainfall can lead to mass movement through three different mechanisms. Often these mechanisms do not act alone. The saturation of soil materials increases the weight of slope materials which then leads to greater gravitational force. Saturation of soil materials can reduce the cohesive bonds between individual soil particles resulting in the reduction of the internal strength of the hillslope. Lastly, the presence of bedding planes in the hillslope material can cause material above a particular plane below ground level to slide along a surface lubricated by percolating moisture.

Earthquakes are another common mechanism that can trigger mass movement. The seismic waves produced by earthquakes vibrate slope materials. This vibration can lead to failure by increasing the downward stress or by decreasing the internal strength of the hillslope sediments through particle movement.

Water, Sediment Transport and Hillslopes

Rainsplash is a microscale process that can be quite effective in moving material on slopes. The impact of rain droplets on the soil surface often detaches individual grains of soil moving them some distance from their source. On flat surfaces, the effect of rain drop impact is to redistribute the material without any net transport in a particular direction. However, on a slope the influence of gravity and slope encourage more material to be redistributed downslope rather than upslope. When slopes become 25 degrees or greater, almost all the redistribution occurs in a downslope direction.

Considerable transport of surface sediments on slopes occurs by rainwash and surface runoff. On relatively flat surfaces, runoff occurs as a continuous layer of water commonly called sheetwash. The erosive potential of sheetwash is usually quite limited because this type of flow is shallow and non-turbulent and cannot readily entrainsurface particles. However, topographic irregularities can quickly transform sheetwash into small channels called rills. Rills then coalesce into larger stream channels and so on. Rills and large stream channels concentrate the movement of water causing an increase in flow velocity and turbulence. Higher flow velocities and turbulence lead to a greater potential for entrainment and subsequent transport of hillslope materials.

Mass Movement in Non-cohesive Materials

Many slopes are composed of non-cohesive, coarse-grained sediments. This type of slope is common to landform features like alluvial fans, screes, talus cones, sand dunes, and glacial outwash deposits. On slopes of this type, mass movement often occurs through the sliding or rolling of a small number of particles as localized instabilities develop. In some cases, these movements can organize themselves into larger avalanches through a domino effect. Mass movement on non-cohesive materials can also occur by way of shallow sliding. Shallow sliding occurs when planes of weakness develop just beneath the surface of the slope. Planes of weakness develop where horizontal layering occurs in the sediment. This layering can be caused by the nature of sediment deposition, percolation of water, or by the presence of subsurface soil, sediment or rock layers.

Scree slope formed of non-cohesive sediments at the base of a steep hill. Mass movement on this type of hillslope occurs primarily by way of localized sliding or rolling of a relatively small number of particles.

In many mountainous areas, shallow coarse soils develop over bedrock. Under the right consitions, large downslope movements of this material can occur. This type of mass movement is known as a debris flow.

Mass Movement in Cohesive Materials

Slopes formed from clays and silts sediments display somewhat unique mass movement processes. Clay and silt particles have a degree of cohesion which gives them potentially more internal strength than non-cohesive sediments. This cohesion occurs because of electrochemical bonds which operate between particles and the surface tension effects of water films that forms around particles. Both of these sources of cohesion are dependent upon moisture content. Maximum cohesiveness takes place when mositure conditions are moderate. Too much or too little water reduces the strength of the cohesion.

The head of a rotational slip in the Black Hills of North Dakota.

Two common types of mass movements in cohesive materials are rotational slips and mudflows. Both of these processes occur over very short time periods. Rotational slips or slumps occur along clearly defined planes of weakness which generally have a concave form beneath the Earth's surface. Rotational slips can be caused by a variety of factors. The most common mechanism reason for them to occur is erosion at the base of the slope which reduces the support for overlying sediments. Erosion at the base of a slope can be caused by the presence of a stream channel or by wave action.

Mudflows occur when slope materials become so saturated that the cohesive bonds between particles is lost. The saturated material then flows like a thick fluid downslope. Flow stops when water loss through seepage causes the sediment to solidify. Mudflows can occur on very low slope angles because internal particle frictional resistance and cohesion is negligible.

Some of the mass movement processes operating on cohesive materials occur over very long time spans. One of the most widespread of these processes is soil creep. Soil creep involves the movement of slope sediments in a series of numerous cyclical steps. The cyclical effects of temperature fluctuations, variations in moisture, and gravity on inclined soil sediments often cause this process.

The Slumgullian mudflow in the San Juan Mountains, Colorado.

Solifluction is the slow movement of soil caused by freeze-thaw action. This process is a widespread in polar and sub-polar regions where permafrost exists. Solifluction occurs when seasonal or daily fluctuations of temperature are above freezing. At these temperatures, the upper portion of the soil surface and permafrost thaw creating waterlogged mass because subsurface ice prevents drainage. The waterlogged mass then flows downslope as lobes of sediment and surface vegetation.

Mass Movement on Hard Rock Slopes

Mass movement on hard rock slopes is often dramatic and quick. Hard rocks derive their internal strength mainly from the strong inter-granular bonds that form when magma cools and crystallizes or when lithification occurs in sedimentary rocks. Because of their strong internal strength hard rock slopes can have relatively steep angles. Nevertheless, weaknesses do occur along bedding planes and joints naturally found in these materials. Most mass movement on hard rock slopes involves the downward movement of small rock fragments pried loose by gravitational stress and freeze-thaw processes. We call these types of mass wasting rockfalls.

Rockfalls from steep hillsides frequently disrupt road traffic in the mountainous regions of British Columbia, Canada.

Larger scale downslope movement of rock can also occur along well defined joints or bedding planes. This type of movement is called a rock slide. Most rock slides occur because the hillslope and rock layers dip in the same direction. In these cases, the rock slide occurs when a fracture plane develops causing overlying materials to slide downslope.

Not all rock slides are the result of the process described above. The rock slide that covered the town of Frank, Alberta, Canada on April 29, 1903 was in part caused by human activities. The sedimentary rock layers in Turtle Mountain dipped away from the valley containing the mining town of Frank. However, joints in the limestone layers did dip toward the town. The mountain was also composed of some structurally weak limestone, shale, siltstone, and coal layers that were deformed by the weight of more massive limestone located above. Finally, the mining of coal at the Turtle Mountain's base reduced the support to overlying materials. Together these factors lead to the sudden movement of 33 million cubic meters of rock in approximately two minutes.

Marine Process

Waves do much of the work of marine processes. They erode, transport and deposit materials. Waves are created by winds as they blow over the surface of the sea. It is the friction between the wind and water that sets waves in motion. The strength of waves depends on the strength of the wind depends on the strength of the wind. It also depends on the length of time and distance over which the wind has been blowing (the fetch).

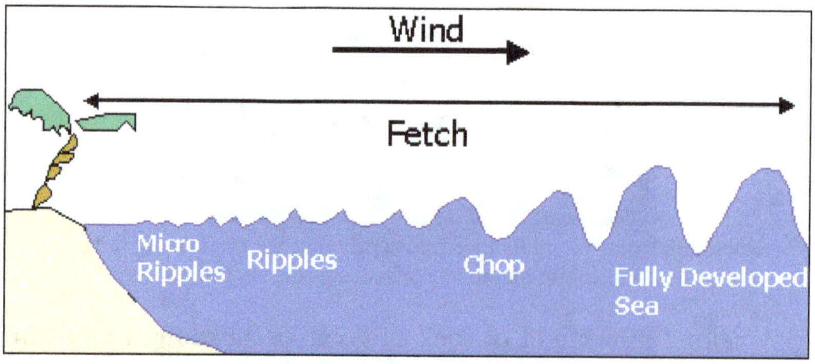

As waves near the coast, they enter into shallower water. Friction with the sea bed causes the wave to tip forward so that it eventually breaks. The resulting forward movement of the water, called the swash, runs up the beach until it runs out of energy. The water then runs back down the beach under gravity. This is called the backwash.

Constructive Waves

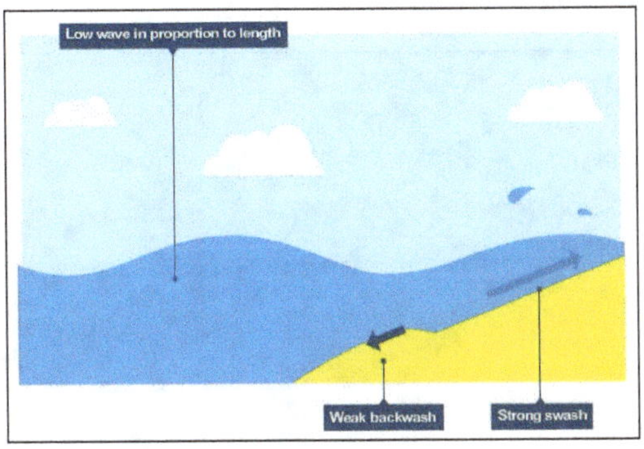

Destructive Waves

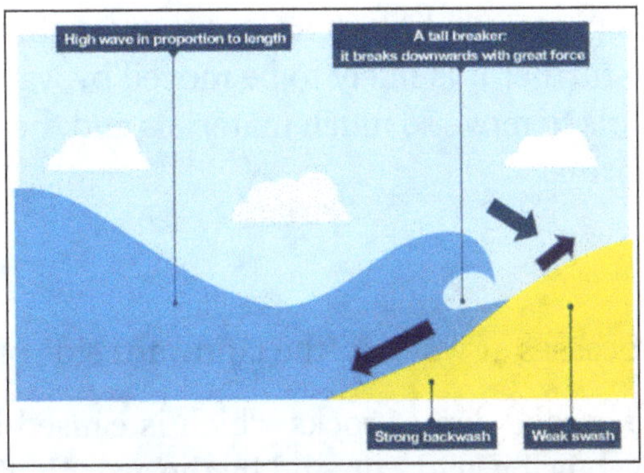

The balance between the swash and the backwash of waves created the difference between constructive and destructive waves. In constructive waves, the swash is stronger than the backwash. As a result, material is moved up the beach and much is left there (deposition). In destructive waves, the backwash is stronger. Material is dragged back down the beach (erosion) and moved along the coast by Long Shore Drift (LSD).

Processes of Erosion

It is destructive waves that do much of the erosion along a coast. They cut away at the coastline in a number of different ways:

- Hydraulic action: This results from the force of the waves hitting the cliffs and forcing pockets of air into cracks and crevices.
- Abrasion: This is caused by waves picking up stones and hurling them at cliffs and thus wearing the cliff away.
- Corrosion: The dissolving of rocks by the sea water.

Attrition is a process whereby the material carried by the waves becomes rounded and smaller over time as it collides with other material. It does not erode the coast as such but does form small pebbles and sand.

Long Shore Drift (LSD)

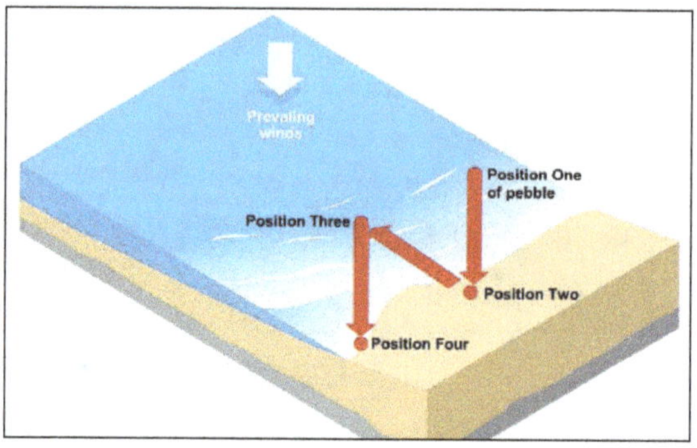

Once rocks are detached from the cliff, waves can move them along the coastline for quite long distances, this process is known as longshore drift. Generally speaking, the smaller the material, the further it is likely to be moved by waves as it is lighter. Eventually, the waves are unable to move so much materials and the materials will be deposited to create new land forms.

Land Processes

There are three main processes at work on the landward side of the coastline:

- Weathering: The breakdown of rocks which is caused by freeze-thaw and the growth of salt crystals, by acid rain and by the growth of vegetation roots.
- Erosion: The weathering away of rocks by wind and rain.
- Mass Movement: The removal of cliff-face material under the influence of gravity in the form of rock falls, slumping and landslides.

Coastal Erosion

Sea erosion at Valiyathura Kerala, India.

Tunnel-like structures formed by erosion in Jinshitan Coastal National Geopark, Dalian, Liaoning Province, China.

Heavy marine erosion: Cliff fall at Hunstanton in the east of England.

Coastal erosion is the loss or displacement of land, or the long-term removal of sediment and rocks along the coastline due to the action of waves, currents, tides, wind-driven water, waterborne ice, or other impacts of storms. The landward retreat of the shoreline can measured and described over a temporal scale of tides, seasons, and other short-term cyclic processes. Coastal erosion may be caused by hydraulic action, abrasion, impact and corrosion by wind and water, and other forces, natural or unnatural.

On non-rocky coasts, coastal erosion results in rock formations in areas where the coastline contains rock layers or fracture zones with varying resistance to erosion. Softer areas become eroded much faster than harder ones, which typically result in landforms such as tunnels, bridges, columns, and pillars. Over time the coast generally evens out. The softer areas fill up with sediment eroded from hard areas, and rock formations are eroded away. Also abrasion commonly happens in areas where there are strong winds, loose sand, and soft rocks. The blowing of millions of sharp sand grains creates a sand-blasting effect. This effect helps to erode, smooth and polish rocks. The definition of abrasion is grinding and wearing away of rock surfaces through the mechanical action of other rock or sand particles.

Small-scale erosion destroys abandoned railroad tracks.

A place where erosion of a cliffed coast has occurred is at Wamberal in the Central Coast region of New South Wales where houses built on top of the cliffs began to collapse into the sea. This is due to waves causing erosion of the primarily sedimentary material on which the buildings foundations sit.

Dunwich, the capital of the English medieval wool trade, disappeared over the period of a few centuries due to redistribution of sediment by waves. Human interference can also increase coastal erosion: Hallsands in Devon, England, was a coastal village washed away over the course of a year, 1917, directly due to earlier dredging of shingle in the bay in front of it.

The California coast, which has soft cliffs of sedimentary rock and is heavily populated, regularly has incidents of housing damage as cliffs erodes. Devil's Slide, Santa Barbara, the coast just north of Ensenada, and Malibu are regularly affected.

The Holderness coastline on the east coast of England, just north of the Humber Estuary, is one of the fastest eroding coastlines in Europe due to its soft clay cliffs and powerful waves. Groynes and other artificial measures to keep it under control has only accelerated the process further down the coast, because longshore drift starves the beaches of sand, leaving them more exposed. The White Cliffs of Dover have also been affected.

Fort Ricasoli in Kalkara, Malta already showing signs of damage where the land is being eroded.

The coastline of North Cove, Washington has been eroding at a rate of over 100 feet per year, earning the area the nickname "Washaway Beach." Much of the original town has collapsed into the ocean. The area is said to be the fastest-eroding shore of the United States' West Coast. Measures were finally taken to slow the erosion, with substantial slowing of the process noted in 2018.

Fort Ricasoli, a historic 17th century fortress in Malta is being threatened by coastal erosion, as it was built on a fault in the headland which is prone to erosion. A small part of one of the bastion walls has already collapsed since the land under it has eroded, and there are cracks in other walls as well.

In El Campello, Spain, the erosion and failure of a Roman farm fish excavated on rock during the first century B.C. was exacerbated by the construction of a close sport harbour.

Hydraulic Action

Hydraulic action occurs when waves striking a cliff face compress air in cracks on the cliff face. This exerts pressure on the surrounding rock, and can progressively splinter and remove pieces. Over time, the cracks can grow, sometimes forming a cave. The splinters fall to the sea bed where they are subjected to further wave action.

Attrition

Attrition occurs when waves cause loose pieces of rock debris (scree) to collide with each other, grinding and chipping each other, progressively becoming smaller, smoother and rounder. Scree also collides with the base of the cliff face, chipping small pieces of rock from the cliff or have a corrasion (abrasion) effect, similar to sandpapering.

Solution

Solution is the process in which acids contained in sea water will dissolve some types of rock such as chalk or limestone.

Abrasion

Abrasion, also known as Corrasion, occurs when waves break on cliff faces and slowly erode it. As the sea pounds cliff faces it also uses the scree from other wave actions to batter and break off pieces of rock from higher up the cliff face which can be used for this same wave action and attrition.

Corrosion

Corrosion or solution/chemical weathering occurs when the sea's pH (anything below pH 7.0) corrodes rocks on a cliff face. Limestone cliff faces, which have a moderately high pH, are particularly affected in this way. Wave action also increases the rate of reaction by removing the reacted material.

Factors that Influence Erosion Rates

Primary Factors

The ability of waves to cause erosion of the cliff face depends on many factors.

The hardness (or inversely, the erodibility) of sea-facing rocks is controlled by the rock strength and the presence of fissures, fractures, and beds of non-cohesive materials such as silt and fine sand.

The rate at which cliff fall debris is removed from the foreshore depends on the power of the waves crossing the beach. This energy must reach a critical level to remove material from the debris lobe. Debris lobes can be very persistent and can take many years to completely disappear. Beaches dissipate wave energy on the foreshore and provide a measure of protection to the adjoining land.

The stability of the foreshore, or its resistance to lowering. Once stable, the foreshore should widen and become more effective at dissipating the wave energy, so that fewer and less powerful waves reach beyond it. The provision of updrift material coming onto the foreshore beneath the cliff helps to ensure a stable beach.

The adjacent bathymetry, or configuration of the seafloor, controls the wave energy arriving at the coast, and can have an important influence on the rate of cliff erosion. Shoals and bars offer protection from wave erosion by causing storm waves to break and dissipate their energy before reaching the shore. Given the dynamic nature of the seafloor, changes in the location of shoals and bars may cause the locus of beach or cliff erosion to change position along the shore.

Coastal erosion has been greatly affected by the rising sea levels globally. There has been great measures of increased coastal erosion on the Eastern seaboard of the United States. Locations such as Florida have noticed increased coastal erosion. In reaction to these increases Florida and its individual counties have increased budgets to replenish the eroded sands that attract visitors to Florida and help support its multibillion-dollar tourism industries.

Coastal Erosion

Pacifica, California coast after major storms in 1997 (resulting from the strongest El Niño on record) destroyed the houses.

Beach erosion at Cabrillo National Monument.

Large-scale coastal erosion at Torrey Pines State Reserve.

Coastal erosion at Torrey Pines State Reserve, resulted in the necessary relocation of a scenic overlook.

Secondary Factors

- Weathering and transport slope processes.
- Slope hydrology.

- Vegetation.
- Cliff foot erosion.
- Cliff foot sediment accumulation.
- Resistance of cliff foot sediment to attrition and transport.
- Human activity.

Tertiary Factors

- Resource extraction.
- Coastal management.

Control Methods

There are three common forms of coastal erosion control methods. These three include: soft-erosion controls, hard-erosion controls, and relocation.

Hard-erosion Controls

A typical seawall that is used for preventing and controlling coastal erosion.

Hard-erosion control methods provide a more permanent solution than soft-erosion control methods. Seawalls and groynes serve as semi-permanent infrastructure. These structures are not immune from normal wear-and-tear and will have to be refurbished or rebuilt. It is estimated the average life span of a seawall is 50–100 years and the average for a groyne is 30–40 years. Because of their relative permanence, it is assumed that these structures can be a final solution to erosion. Seawalls can also deprive public access to the beach and drastically alter the natural state of the beach. Groynes also drastically alter the natural state of the beach. Some claim that groynes could reduce the interval between beach nourishment projects though they are not seen as a solution to beach nourishment. Other criticisms of seawalls are that they can be expensive, difficult to maintain, and can sometimes cause further damage to the beach if built improperly.

Natural forms of hard-erosion control include planting or maintaining native vegetation, such as mangrove forests and coral reefs.

Soft-erosion Controls

Sandbagged beach at the site of Hurricane Sandy.

Soft erosion strategies refer to temporary options of slowing the effects of erosion. These options, including Sandbag and beach nourishment, are not intended to be long term solutions or permanent solutions. Another method, beach scraping or beach bulldozing allows for the creation of an artificial dune in front of a building or as means of preserving a building foundation. However, there is a U.S. federal moratorium on beach bulldozing during turtle nesting season, 1 May – 15 November. One of the most common methods of soft erosion control is beach nourishment projects. These projects involve dredging sand and moving it to the beaches as a means of reestablishing the sand lost due to erosion. In some situations, beach nourishment is not a suitable measure to take for erosion control, such as in areas with sand sinks or frequent and large storms.

Relocation

Relocation of infrastructure any housing farther away from the coast is also called managed retreat. The natural processes of both absolute and relative sea level rise and erosion are considered in rebuilding. Depending on factors such as the severity of the erosion, as well as the natural landscape of the property, relocation could simply mean moving inland by a short distance or relocation can be to completely remove improvements from an area. Typically, there has been low public support for "retreating."

Tracking

Storms can cause erosion hundreds of times faster than normal weather. Before-and-after comparisons can be made using data gathered by manual surveying, laser altimeter, or a GPS unit mounted on an ATV. Remote sensing data such as Landsat scenes can be used for large scale and multi year assessments of coastal erosion.

Biogeomorphology of Coastal Systems

The biogeomorpholgy of both hard and soft substratum including the role in which plants and animals play in stabilizing and de-stabilizing these aquatic systems is also explored.

Biogeomorphology considers the interactions between the ecology and geomorphology of a system. Ecology is the study of relationships between the biota and their environment, and geomorphology examines landforms and how they are formed. Biogeomorphology can be studied in terrestrial as well as aquatic systems. Within aquatic systems biogeomorphological relationships can be found for both hard substrates (rocky shores/coral reefs) and soft substrates (muddy/sandy coastal sediments). Biota can alter geomorphology by creating hard substrates (e.g. coral reefs) or by modifying the stability or erodability of soft substrates. Key species in the benthic communities of sediment shores can influence geomorphology by acting as biostabilizers or biodestabilizers of sediment. The impact of organisms on coastal morphology can be both dramatic (e.g. coral reefs, saltmarshes, mussel beds) and more subtle by modifying rates of sediment erosion and accretion. Biota are not only influenced by various environmental factors (abiotic – physical and chemical; biotic – other organisms; anthropogenic – humans), but organisms also act as ecosystem engineers and modify their environment (physical, chemical and biological aspects). Ecosystem engineers create valuable habitats (e.g. coral reefs and saltmarshes) that are important for fish and birds and serve as a vital part of our natural coastal defence against storm surges and sea level rise.

The term biogeomorphology was first used in the eighties, although earlier studies focused on the topic without using this term. In coastal systems biogeomorphological interactions are clearly demonstrated in shallow, productive waters and various sedimentary environments. Examples of biogeomorphological interrelationships include sand dune development, tidal flats, salt marshes, mangrove systems and coral reefs.

Relevant geomorphological factors in coastal systems are bathymetry, bed composition (rock, gravel, sand, silt), and the transport of sediment. It also includes factors that drive morphological processes, such as water flow and wave energy. The biota involved in coastal biogeomorphologyinclude plants and animals, ranging from very small (microorganisms) to very large (feeding pits of whales).

The geomorphological directly affects biota through its influence on habitats. Coastal morphology and geomorphological processes define the gradients between high and low, between wet and dry and between sedimentation and erosion. These gradients and the processes that cause them determine the gradients in grain size, nutrient levels, organic matter levels and water content. Plants and animals are adapted to specific conditions and will therefore be found and abundant at specific locations (often forming habitats).

The biological influence on geomorphological processes is through the ability of biota to create, maintain or transform their geomorphological surroundings. This is demonstrated by the influence of vegetation and macrofauna on water flow, sediment erosion and deposition, or by the influence of fauna on sediment characteristcs through bioturbation and biostabilization.

In some cases morphological processes are dominant over biological processes and therefore it is necessary for biota to adjust to their environment. In other cases biological processes are dominant (e.g. coral reefs, salt marshes). The most interesting are those cases where there is a mutual interaction that leads to feedback coupling of processes. When looking for these cases, it is important to examine the spatial and temporal scales of the mutually interacting processes.

Worm casts produced by Arenicola marina on sandy/muddy beach increase bed roughness and sediment resuspension.

Biogeomorphology for Hard Substrates

On rocky shores and coral reefs the typical community of organisms modifies the erosion of substrate. Influenced by abiotic factors such as wave energy, splash water, inundation frequency and period, depth, desiccation and substrate type, a clear zonation of various cyanobacteria, macro-algae, fungi, lichens, molluscs, sponges, worms, sea urchins, fish, etc. can be found. Some of these organisms dwell on the surface of the substrate, while others live within the substrate. Their effect on erosion of the substrate is divided into 'biological corrosion' processes that modify the substrate but produce no erosion product, and 'biological abrasion' processes that generate an erosion product. Grazing, burrowing and boring on or in the substrate involves biological abrasion, and is most apparent in coral reef systems.

Biogeomorphology for Soft Substrates

In soft coastal systems, the interrelationships between geomorphological factors and biota mainly apply to benthic fauna and flora. The presence of benthic species is affected by hydraulic and morphologic conditions, such as water depth or elevation on the shore, current velocity, wave action, salinity and grain size. The biota responsible for changes in geomorphology of soft substrates can be divided into two functional groups, namely biostabilizers and biodestabilizers. Biostabilization leads to increased sediment stability and a reduction in erosion potential, whereas biodestabilization leads to reduced sediment stability and an increase in erodibility.

Biostabilization by Plants

On tidal flats, small algae (diatoms) are capable of affecting the geomorphology. These diatoms can form extensive algal mats and excrete extracellular polymeric substances, which is a sticky mucus-like substance made of polysaccharides that glues the sediment together and therefore protects the sediment against erosion.

Seagrass meadow.

Seagrass is dependent on clear water, it needs sunlight to grow. A seagrass meadow slows down the current velocity near the bed, therefore reducing the resuspension of sand and silt, and this in turn helps to maintain clear water. Their root system also promotes binding of the substrate. In addition, seagrass meadows encourage deposition of suspended sediment, which leads to the supply of organic material and nutrients needed for growth.

Seaweeds and salt marsh plants are capable of modifying their physical environments by damping down wave energy and tidal currents. Salt marsh vegetation can also promote sediment deposition resulting in a gradual elevation of the marsh on the upper shore. The higher the marsh gets, the more vegetation can grow and the better the protection against erosion. Other stabilizing effects result from cementation of beach-rock by cyanobacteria and stromatolite formation by algae. Recent research of Temmerman et al. points out that the interaction of vegetation growth, tides and sedimentation is crucial in the formation of the typical levee-basin configuration of the saltmarsh landscape.

Upper shore salt marsh with creek system and evidence of benthic diatoms stabilizing mudbanks.

Biostabilization by Animals

Example of biostabilization by animals.

Some macrozoobenthos can actively filter sediment particles from the water column and deposit them on the bed. The presence of a mussel bank, for example, will alter the bed in different ways. Mussels at high densities will protect the underlying sediment from erosion by waves and currents (armouring effect). In addition, mussels are filter feeders and actively remove small suspended particles from the water column and any inorganic material (e.g. silt) is rejected as faeces and pseudofaeces. Some of this material will get deposited and incorporated into the bed, causing a shift in the composition to finer sediments and an increase in bed-level. Animal tube fields at high densities can also stabilize the sediment, because there is an accumulation of fine particles and organic matter between the tubes. The tubes may also reduce near-bed flow and create skimming flow above thus creating a stabilizing effect. Mucus production by the community of microorganisms, meiofauna and macrofauna between the tubes may also contribute to sediment stability.

Biodestabilization

Benthic fauna may destabilize the substrate by their burrowing and surface deposit feeding activity. They also destabilize the sediment as a result of grazing on biostabilizers such as microalgae. The constant mixing and recycling of sediment in the top centimeters of the bed, known as bioturbation, results in a change in the vertical particle size profile. Selective uptake of preferred particle sizes and their subsequent defaecation results in sorting and pelletizing of sediments. Together, the burrowing and the constant movement within the substrate, results in the generation of a surface micro-relief that has a higher hydraulic roughness, which increases turbulence and increases the potential for erosion. Furthermore, bioturbation also affects the sediment water content, porosity and sediment cohesion. Research of Paarlberg et al. has shown that impacts of destabilizing biota on mudflat morphology and bed composition can be significant and can be quantified by adding a limited number of algorithms to existing hydro-morphological models. Borsje et al. show that the influence of (de)stabilizers might extend beyond the mudflat scale to the estuary scale in cases of meso-tidal highly productive systems, such as the Wadden Sea.

Example of bioturbation.

Zoning

Examples of biota acting as ecosystem engineers in the intertidal. Key biota are divided into two functional groups, the bio-stabilisers and bio-destabilisers, with varying spatial distribution along estuarine gradients (both axial and vertical).

Coastal Processes and Beaches

Waves, tide, and wind dominate coastal processes and landforms. Rivers deliver sediment to the coast, where it can be reworked to form deltas, beaches, dunes, and barrier islands.

Coastal Processes

The coastal zone is that part of the land surface influenced by marine processes. It extends from the landward limit of tides, waves, and wind blown coastal dunes, and seaward to the point at which waves interact significantly with the seabed.The coastal zone is a dynamic part of the Earth's surface where both marine and atmospheric processes produce rocky coasts, as well as beaches and dunes, barriers and tidal inlets, and shape deltas. The atmospheric processes include temperature, precipitation, and winds, while the major marine processes are waves and tides, together with water temperature and salinity. The coast also supports rich ecosystems, including salt marshes, mangroves, seagrass, and coral reefs. The diverse coastal ecology is favored by the shallow waters, abundant sunlight, terrestrial and marine nutrients, tidal and wave flushing, and a range of habitat types.

Waves — Generation and Types

Waves provide about half the energy to do work at the coast. Ocean waves are generated by wind blowing over the ocean surface. The stronger the wind, the longer it blows and the longer the fetch, or stretch of ocean over which it blows, the larger the waves. The world's greatest wave factories are in the zone of sub-polar lows centered on 40–60° N and S latitudes, the so-called roaring 40's and screaming 60's. The strong westerly winds produce the world's biggest waves which initially head west, and are deflected equatorward by the Coriolis effect, arriving from the northwest in the northern hemisphere and southwest in the southern hemisphere. Other major wave climates are the easterly waves produced by the expansive but moderate velocity northeast and southeast Trade winds — and lesser seasonal waves produced by the monsoons and even the polar easterlies, together with occasional hurricanes that can produce massive waves as well as storm surges.

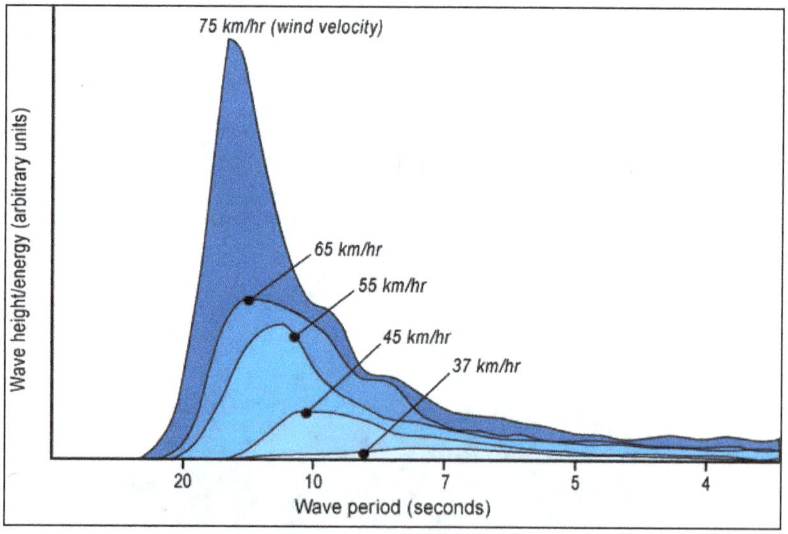

As wind velocity increases, the period or time between waves, and wave length, increases, and the amount of energy transferred to the waves increases exponentially. Note how as wind velocity doubles from 37 to 75 km/hr the amount of energy increases exponentially. Very strong winds are therefore required to generate the biggest waves.

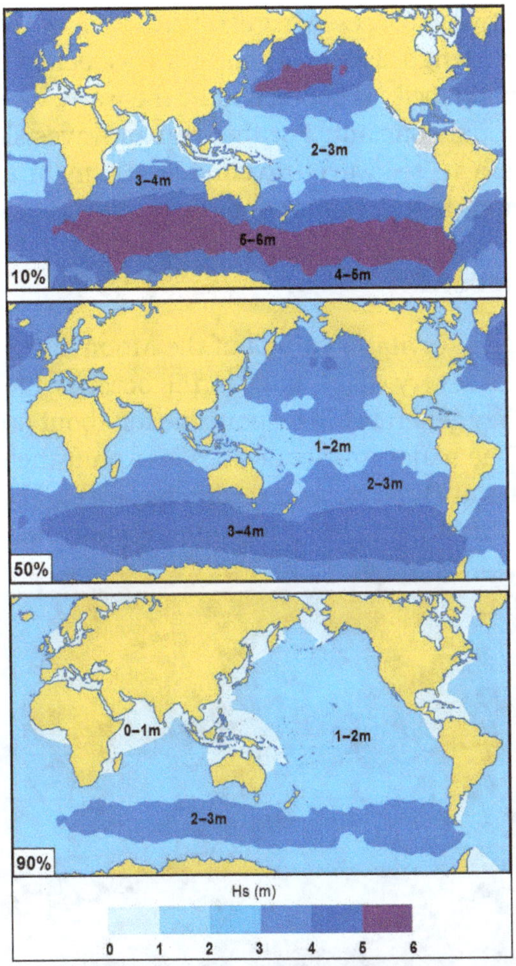

The highest waves occur in the Southern Ocean and north Pacific and Atlantic where they are generated by strong sub-polar lows. They average 5–6 m in height 10% of the year (top). The Southern Ocean has the most persistent higher waves averaging 3–4 m 50% of the time (centre), and at least 2–3 m 90% of the year (bottom). While the sub-tropical trade winds also blow year round across huge sections of ocean, they are of only moderate velocity, and hence generate only moderate waves.

When waves are being generated they are called a sea and consist of short, steep, high, slower waves, which tend to topple over and break, and have a broad spectrum of direction. Once the wind stops blowing, and the waves leave the area of generation, they quickly transform into swell – lower, longer, faster and uniform in direction. Swell waves can theoretically travel around the world with minimal loss of energy, while in reality they eventually break on some distant shore.

Waves are a form of potential energy that can be transported across hundreds to thousands of kilometers of ocean to be released as kinetic energy when they shoal and break. Waves are defined by their height (H) (trough to crest), length (L) (crest to crest) and period (T) (time between successive crests). The longer the period the longer and faster the waves, as wave length L=1.56 T, and wave velocity $C = 1.56\ T^2$. When waves enter shallow water their velocity is controlled by the water depth (d) such that $C=\sqrt{gd}$ where g is the gravitational constant. For this reason waves slow down as they move towards the shore, with a 10 sec wave traveling at 56 km/hr (35 mph) in deep water slowed to 7 km/hr (4 mph) in 5 m water depth.

Tides and Tidal Currents

Tides are produced by the gravitational pull of the Moon and Sun acting on a rotating Earth. This pull produces a very slight bulge in the ocean, which we know as tide. The tides and the currents they generate are responsible for about 50% of the marine energy delivered to the coast. The major impact of tides is to shift the shoreline between high and low tide, and to generate tidal currents either parallel to the coast, or at tidal inlets and estuaries, currents flowing into the inlets and perpendicular to the coast.

Tidal inlet at Merrimbula, Australia. Tidal currents ebb and flood though the narrow inlet moving marine sand from the beach into the inlet to form a deep tidal channel and shallow sandy tidal delta.

Wind and Currents

Winds blowing over the oceans are responsible for generating ocean waves. Nearer the coast they can generate local seas – they can move the ocean surface and generate locally wind driven currents which in places can result in upwelling and downwelling. Finally, when blowing over the beach, they can transport sand inland to build coastal sand dunes.

Fluvial-deltaic Systems

Fluvial systems deliver sediment to the coast where it is deposited in estuaries and deltas. Depending on their location, deltas are also acted on by waves, tides, and other currents, and shaped to suit the prevailing processes. Sediment can also be moved longshore to supply beach and barrier systems.

The Gasgoyne River delta in Western Australia delivers large volumes of sand to the coast where it is deposited in river mouth shoals and slowly reworked longshore to supply downdrift spits, barriers, and dunes.

Sea Level

Sea level determines the position of the shoreline. During the last glacial maxima (ice age) 18,000 years ago, sea level was 120 m below present, and the continental shelves were exposed. It then rose, reaching present sea level around 6,000 years ago, after which it was relatively stable. Now, with climate change, it is beginning to rise again, and may rise as much as 1 m over the next 100 years, triggering shoreline retreat, inundation, and erosion.

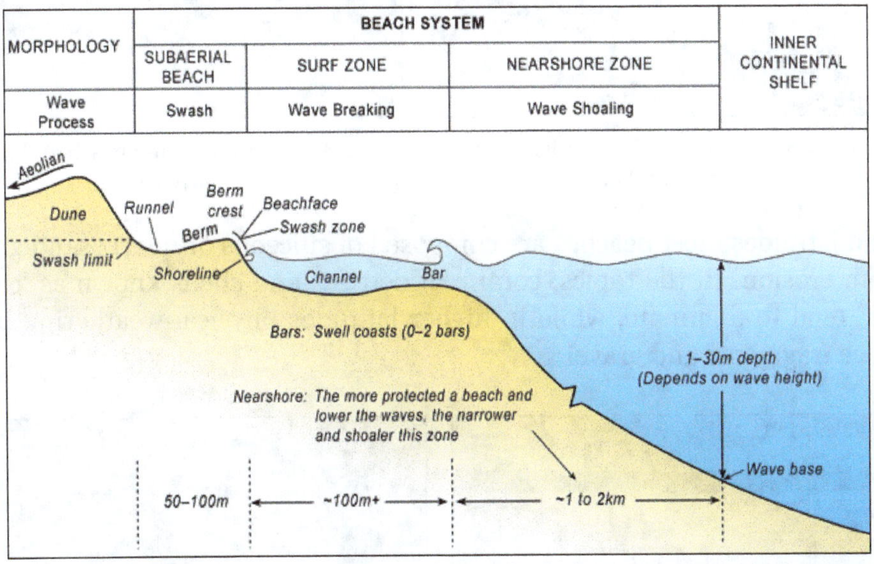

An idealised cross-section of a wave-dominated beach system consisting of the swash zone which contains the subaerial or 'dry' beach (runnel, berm, and beach face) and is dominated by swash processes; the energetic surf zone (bars and channels) with its breaking waves and surf zone currents; and the nearshore zone extending out to wave base where waves shoal building a concave upward slope.

Beaches are wave-deposited accumulations of sediment located at the shoreline. They require a base to reside on, usually the bedrock geology, waves to shape them, sedi-

sediment to form them, and most are also affected by tides. The beach extends from wave base where waves begin to feel bottom and shoal, across the nearshore zone, though the surf zone to the upper limit of wave swash. In the coastal zone ocean waves are transformed by shoaling, breaking, and swash. In doing so they interact with the sea-bed, and determine the beach morphology or shape, a process called beach morphodynamics.

Beach Sediment – Types and Sources

All beaches consist of sediment, which can range in size from sand up to cobbles and boulders. The finer sand result in very low gradient (~1°) beaches while cobbles may be stacked as steep as 20°. Most beaches with fine to medium sand have a swash zone gradient between 1–8°.

A steep cobble-boulder beach at Catfish Bay, Western Australia.

Low-gradient fine sand beach at Ball Bay, Queensland.

In the mid latitudes most beaches are composed of siliceous or quartz sand grains derived from erosion. In the topics, coral reef detritus and shells known as 'carbonate sediment' tend to dominate, while in higher latitudes physical weathering produces coarse rock fragments and gravel.

Shell Beach, Western Australia is composed entirely of the white cockle shell Fragum erugatum.

Close up view of the cockle shell.

Greenmount Beach in Queensland, widened by 200 m as a result of beach nourishment.

Collaroy Beach in NSW has been eroding, resulting in the construction of rock seawalls.

Beach Sub-systems

At the beach the three zones of wave transformation (shoaling, breaking, and swash) produce three morphologically distinct sub-systems. The shoaling wave zone builds a low gradient (~1°) concave upward profile, smooth in outline, with small wave-generated ripples and generally onshore sediment movement. As they shoal they interact with the seabed, slowing down and increasing in steepness and height.

View of Makapu Beach, Hawaii, showing waves shoaling and steepening as they travel across and interact with the nearshore zone, then breaking across the surf zone.

The surf zone is the most dynamic part of the beach and extends from the breaker zone to the shore. Waves break when the water depth is approximately 1.5 times the wave height. They can break as a spilling breaker on low gradient slopes, a plunging wave on moderate gradients, or a surging wave on steep slopes. In breaking, waves transform their potential energy to kinetic energy, which is initially manifest as the broken wave of translation, or wave bore, which moves shoreward as broken white water. At the shoreline the currents can be deflected longshore and water may return seaward as a rip current.

Surf zone currents can transport sediment onshore, longshore and offshore and build the (sand) bars and troughs that occupy the surf zone. The number and location of the bars is a product of infragravity waves, a low frequency (greater 30 sec period) wave produced by sets of higher and lower waves and which is enhanced by wave breaking across the surf zone. The longer the infragravity wave period the more widely spaced the bar(s). Another form of infragravity wave called edge waves also influence the longshore spacing of rip currents and channels, which are typically 200–300 m apart on ocean beaches. Rip currents are narrow, seaward moving currents that move seaward though the surf zone, often in a deeper rip channel. They are a mechanism for returning the water back out to sea, and a conduit to transport seaward eroded beach sediment during high seas. They are also a major hazard to beach goers and responsible for most beach rescues and drowning.

Low waves breaking on a shallow bar and flowing shoreward into a rip feeder channel. The dye highlights the rip feeder current flowing along the base of the beach face, then turning to flow seaward in the deeper rip channel.

A rip current (center) carrying suspended sediment flowing out beyond the surf zone as a rip head, with an earlier rip head meandering to the right, Miles Beach.

When the broken wave reaches the base of the wet beach it collapses and runs up the beach face as swash or uprush in the swash zone. The uprush stops toward the top of the slope, some percolates into the beach, the remainder flows back down the beach as backwash. Both actions produce a relatively steep seaward sloping swash zone or beach face, which can range from 1 to 20°. As sediment is deposited in the swash zone it can build a berm, a near horizontal to slightly landward-dipping sand surface, the area where most people sit when they go to the beach. The swash zone may also contain beach cusps, spaced about every 20 to 30 m and produced by another form of edge wave.

Wave runup on the steep beach face at Ke lli Beach.

A steep reflective beach with well developed high tide beach cusps at Hammer Head.

Beach Types

Beaches can range from low energy systems, where small waves lap against the shore, to those with high waves breaking across several hundred meters of surf zone. They can also be exposed to micro (<2 m), meso (2–4 m), macro (4–8 m) and mega tides (>8 m). To accommodate the range of wave heights, tide ranges and beach sediment (sand to boulders) beaches are divided into the three basic types: wave-dominated, tide-modified, and tide-dominated, based on the relative tide range (RTR).

$$RTR = TR/H$$

where TR is the spring tide range.

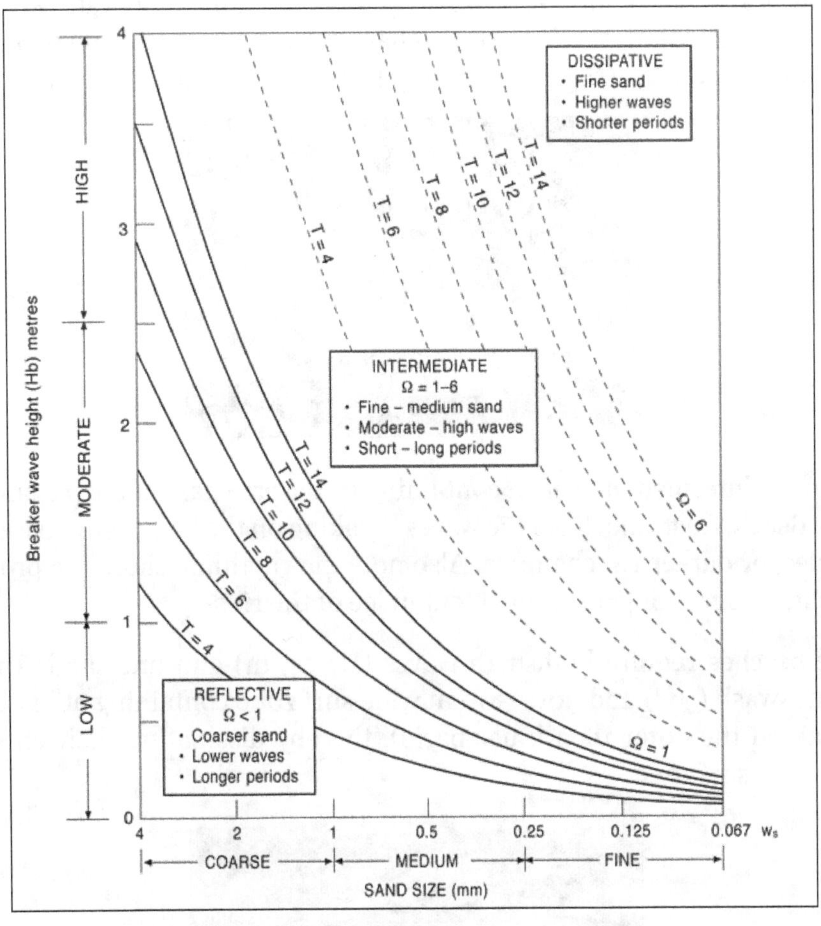

Wave-dominated beaches have an RTR tide range less than three times the average wave height (RTR < 3). They consist of three states – reflective, intermediate, and dissipative. Reflective beaches are produced by lower waves (H<0.5 m), longer wave periods, and coarser sediment, such that all coarse sand and cobble/boulder beaches are reflective. They consist of a relatively steep beach face (5–20°), with waves breaking by surging at the base of the beach and running up the beach face and returning as backwash to 'reflect' off the beach face. There is no bar or surf zone.

A plot of breaker wave height versus sand size, together with wave period, that can be used to determine the approximate beach state for wave-dominated beaches. To use the chart, determine the breaker wave height, period and grain size (mm). Read off the wave height and sand size, then use the period to determine where the boundary of reflective/intermediate, or intermediate/dissipative beaches lies. $\Omega = 1$ along solid T lines and 6 along dashed T lines. Below the solid lines $\Omega < 1$ and the beach is reflective; above the dashed lines $\Omega > 6$ and the beach is dissipative; between the solid and dashed lines Ω is between 1 and 6 and the beach is intermediate. ($\Omega = H_b/W_sT$ and is known as the dimensionless fall velocity).

Intermediate beaches are common on open coasts and require moderate waves (H=0.5–2.5 m) and fine to medium sand. They are characterized by a surf zone with one or two bars up to 100 m wide. The bar is usually cut by regular rip channels and currents.

Well-developed intermediate beach containing transverse bars and rip channels along Lighthouse Beach, Australia. Note the waves breaking on the bars, with no waves breaking in the deeper darker rip channels. Also note the rhythmic shoreline protruding in lee of the bars and forming an embayment in lee of the rips.

Dissipative beaches require both high waves (H>2.5 m) and fine sand. They have a low gradient swash (~1°) and 300–500 m wide surf zone containing at least two bars. Waves break on the outer then inner bar(s), thereby dissipating their energy as the move across the surf zone

Long, high Southern Ocean swell arriving at Dog Fence Beach, South Australia. The swell breaks over the wide outer bar, reforms in the central trough, then breaks across the inner bar, resulting up to 10 lines of breakers and a 500 m wide dissipative beach and surf zone.

Tide-modified beaches have an RTR between 3 and 10, implying tide range is increasing and wave height decreasing. They usually have a steep, coarser-grained, cusped, reflective, high tide beach. This is fronted by a wide, finer-grained, low gradient, often featureless, intertidal zone, up to 200 m wide, then a low tide surf zone which may contain bars and rip channels.

A steep reflective high tide beach face fronted by a 100 m wide tide-modified low tide terrace crossed by shallow drainage channels at North Harbour Beach.

Tide-dominated beaches have an RTR between 10 and approximately 50, implying high tides and very low waves ($H<<0.3$ m). They consist of a low elevation, coarse-grained, irregular, high tide beach, fronted by low gradient ($<<1°$) inter-tidal sand, and even mud flats that may be hundreds of meters wide. Beyond an RTR of 50, tidal flats prevail.

A narrow high tide beach fronted by 1 km wide inter-tidal sand flats, upper Spencer Gulf.

An additional beach type consists of a high tide reflective beach face fronted by inter-tidal rocks flats, and in the tropics a high tide beach fronted by a fringing coral reef flat. Furthermore any beach located in the high latitudes will be seasonally exposed to freezing air and water temperatures leading to the development of sea ice, shoreface ice, and a frozen snow covered beach.

A high tide beach fringed by coral reef at Turquoise Bay, part of the Ningaloo Marine Park.

During summer, with floating ice against the shore.

During freeze-up, with snow and sea ice accumulating.

The frozen winter beach and ocean.

Beaches and Barriers

A coastal sand barrier consisting of a beach and vegetated dunes, backed by a lagoon, at Big Beach, Queensland.

Beach systems are an essential component of a larger scale coastal landform called barriers, which are long-term accumulation of wave, tide, and wind deposited marine sediment (usually sand) at the shore. They consist of a beach from which sand is transported either onshore, via wave overwash and wind blown sand, or alongshore, to form spits and move into tidal inlets. When separated from the mainland by lagoons and marshes they are called barrier islands, which occur along the US East and Gulf coasts. Some are backed by large dune systems as along the Oregon coast.

References

- 4-igneous-processes-and-volcanoes: opengeology.org, Retrieved 17 May, 2020
- Marine-processes, section-a-coastal-environments: maxwatsongeography.wordpress.com, Retrieved 08 June, 2020
- Valvo, Lisa M.; Murray, A. Brad; Ashton, Andrew (1 June 2006). "How does underlying geology affect coastline change? An initial modeling investigation". Journal of Geophysical Research: Earth Surface. 111 (F2): F02025. doi:10.1029/2005JF000340
- Biogeomorphology-of-coastal-systems: coastalwiki.org, Retrieved 15 August, 2020
- Orme, Anthony R. (2007). "The Rise and Fall of the Davisian Cycle of Erosion: Prelude, Fugue, Coda, and Sequel". Physical Geography. 28 (6): 474–506. doi:10.2747/0272-3646.28.6.474
- Coastal-processes-and-beaches-26276621: nature.com, Retrieved 19 April, 2020
- Whittaker, Alexander C. (2012). "How do landscapes record tectonics and climate?". Lithosphere. 4 (2): 160–164. Bibcode:2012Lsphe...4..160W. doi:10.1130/RF.L003.1

Weathering and Erosion

4

- Weathering
- Physical Weathering
- Chemical Weathering
- Honeycomb Weathering
- Frost Weathering
- Erosion
- Types of Erosion
- Consequences of Human-made Soil Erosion
- Preventing and Managing Erosion

The breaking down of rocks, soil and minerals when they come in contact with the Earth's atmosphere, water and organisms is defined as weathering. Erosion refers to the act of eroding the surface by wind, water and other natural agents. This chapter sheds light on weathering and erosion for providing an in-depth understanding of the subject.

Weathering

The Earth's lithosphere is made up of varieties of rocks and relief features. Landforms are the most widespread geological and geographic phenomena on earth. Their origin, dynamism and distribution are studied under the branch of geomorphology. The Earth's crust is divided into two types as Continental Crust and Oceanic Crust. Mountains, hills, rock exposures and stony wastelands all show a lot of relief features on continental regions. The surface of the Continental Crust is under the influence of Sun's radiation, atmospheric temperature variation and the action of water and other fluids. Earth's surface relief features are the reflections of the geology and structure.

The surface relief of any land is not permanent on the earth. Every block of rock is subjected to modifications. Over a period of geological time, these surface blocks and rocks break down into smaller and finer pieces. Since the rocks present in the uppermost layers of the earth's crust are in close interaction with the atmosphere, hydrosphere and biosphere, they are not stable.

Sun's radiant energy, role of water from the hydrological cycle, oxygen and carbon-di-oxide of the atmosphere and the action of organic acids over the rocks, play a significant role in the material transformation over the surface of the earth. All igneous and metamorphic rocks exposed at or near the surface of the earth are in an unstable condition. They are unstable both physically and chemically. Geomorphic processes are all those physical and chemical changes which effect a modification of the earth's surficial form. There are three distinct degradational processes which affect the land regularly. They are weathering, mass-wasting and erosion.

The physical disintegration and chemical decomposition of a rock on land, is known as weathering. It is a unique phenomena on the earth' surface:

- Factors influencing Weathering.
- Physical Weathering Processes.
- Chemical Weathering Processes.
- Biological Weathering Processes.
- Rate of Weathering.

Factors Influencing Weathering

Weathering is a collective term for the mechanical, chemical and biological (organic) processes that take place on the rock bodies of the earth's surface. In simple terms, weathering is defined as the natural breakdown of rocks into minor fragments, soils and sediments.

The impacts of Weathering are very phenomenal. The nature and magnitude of weathering differs from place to place. The first impact is the physical modification of pre-existing rocks. Several modifications are made on the pre-existing rocks by mechanical forces, chemical reactions and biological interactions.

These changes are expected to take place in the existing geologic materials exposed at or near the surface of the Earth. The second impact is destabilization. Weathering destabilizes the surface materials and encourages removal by erosion. The third impact is soil formation. Soils are products of weathering. Weathering creates both soils and loose rock fragments. The fourth impact is the chemical modification.

Weathering releases chemical compounds to support many biological processes. Weathering is of two types as:

- Physical weathering (which is also called as mechanical weathering).
- Chemical weathering.

Mechanical weathering is related to the physical breakup of rocks into small pieces and fragments. There are many factors which influence the weathering of minerals and rocks. They are:

- Topography.
- Climatic conditions: Temperature and humidity.
- Physical characteristics of rocks.
- Chemical and structural characteristics of rocks: Mineralogy, structural features.
- Vegetation: Abundance and type micro and macro vegetation.

Topography is an important factor in relating rocks with the atmospheric pressure, temperature and water vapor. The climatic condition tends to control the kind and rate of weathering. Under conditions of low rainfall, there is a dominance of physical weathering which reduces the size and increases the surface area with little change in volume.

The increase in moisture content encourages both chemical as well as mechanical changes. This also creates new minerals and soluble products. The weathering rates are generally fastest in humid tropical regions as there is sufficient moisture and warmth to encourage chemical decomposition. The easily weatherable minerals disappear on account of intense chemical weathering and more resistant products (hydrous oxides of Fe and Al) tend to accumulate. Climate controls the dominant type of vegetation which in turn controls the biochemical reactions in soils and mineral weathering.

Physical characteristics also influence the process of weathering. The characteristics are:

- Differential composition.
- Particle size.
- Hardness and degree of cementation.

Chemical and structural characteristics of minerals present in rocks are also expected to play a significant role in weathering. For minerals of given particle size, chemical and crystalline characteristics determine the ease of decomposition. (e.g.) gypsum – sparingly soluble in water, is dissolved and removed in solution form under high rainfall.

Ferro magnesium minerals are more susceptible to chemical weathering than feldspar and quartz Tightness of packing of ions in crystals is yet another property for encouraging mineral weathering processes. Less tightly packed minerals like olivine and biotitic are easily weathered as compared to tightly packed zircon and muscovite (resistant). The structures of rocks are fully encouragive to weathering activities. Presence of joints in rock bodies are typical examples.

Physical Weathering Processes

In physical weathering, there is no change in the chemistry of the parent rock. The primary process of physical weathering is abrasion. It is the process in which rock blocks are reduced in their sizes. Sand and other particles normally move above rock bodies. The mechanical scraping of a rock surface by friction between rocks and these moving particles is known as abrasion. Particles are transported by wind, glacier, waves, gravity, running water or erosion.

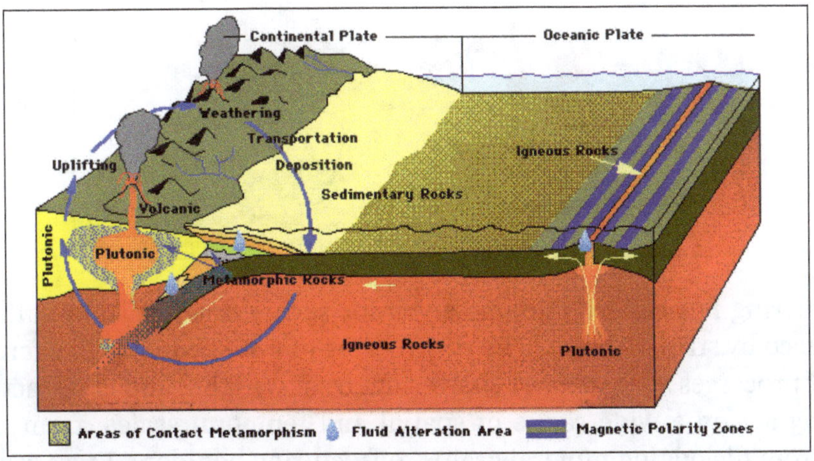

After friction, the moving particles dislodge the loose and weak debris from the sides of the rock. These particles can be dissolved in the water and taken away to some other location. The intensity of abrasion depends on the hardness, concentration, velocity and mass of the moving particles. The major incidences responsible for physical weathering are expansion resulting from unloading, crystal growth, thermal expansion, organic activity and colloidal plucking.

The mechanisms of physical weathering include:

- Freezing and thawing.
- Root wedging.
- Heat Spalling.

Recurring hotness and coldness is an important controlling factor in weathering. One of the main causes of physical weathering is the formation of ice in cracks or cavities within rocks.

Freezing and thawing is the first set of processes: First, water soaks into the cracks or cavities, of the rocks below the surface and gets sealed up. Then, if the temperature falls low enough, the water freezes. When water freezes into ice, its volume increases by nine percent. As a result of this, it expands (thaws) in the cracks and may push hard enough to split the rock into minor fragments. Under specific circumstances, this expansion is able to displace and create fractures in the rocks. Not all volumetric expansion is caused by the pressure of the freezing water; it can be caused by stresses in water that remains unfrozen.

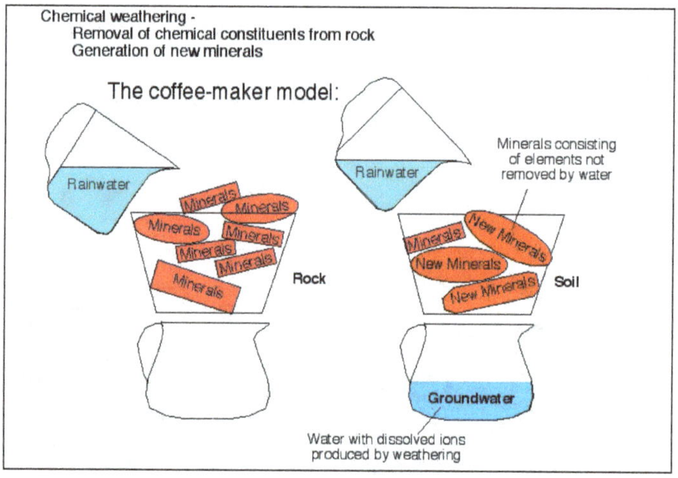

Frost weathering is a collective name given for several mechanical weathering processes induced by stresses created by the freezing of water into ice. The term includes a variety of processes such as frost shattering, frost wedging and cryofracturing. The process may act on a wide range of spatial and temporal scales, from minutes to years and from dislodging mineral grains to fracturing boulders. Frost weathering is mainly driven by the frequency and intensity of freeze-thaw cycles and the properties of the materials subject to weathering. It is most pronounced in high altitude and latitude areas.

The second mechanism is the root wedging. Plant root, have a great role to play in weathering. Plants are effective agents of mechanical weathering. Roots can penetrate through the cracks of rocks to depths of several meters. As the roots grow, they exert a tremendous amount of pressure on the walls of the cracks. This breaks them into pieces. Root wedging is a major process of mountains containing forests and natural vegetation.

Heat spalling is another process of mechanical weathering. Heat from forest fires will cause the outer surface layers of rock to expand quickly and break away in spalls.

There are two more processes of physical weathering as exfoliation and spheroidal weathering:

- Exfoliation denotes the loss of outer layers of rock bodies as it weathers and detaches from the main mass. This leaves a dome like structure. Exfoliation domes are typical geomorphic features.
- Spheroidal weathering is a unique process. In this type of weathering, the angular rock fragments weather into spherical shapes due to the increased vulnerability of edges and corners to attack.

Chemical Weathering Processes

The second major type of weathering is the Chemical Weathering. It is the process of decomposition of Earth's surface materials. It is done by various agents and chemical reaction. The reactions, here, are exothermic and result in the increase in volume of minerals that contribute to the physical disruption of rock.

The effectiveness of chemical weathering is directly related to the:

- Mineral stability in weathering.
- Amount of surface area exposed, which itself is exponentially related to the.
- Density of fractures in a rock.

Most chemical weathering results in:

- An increase in bulk with resulting strains and stresses within the rocks.
- Lower density minerals.
- Smaller particle size and increased surface area.
- More mobile minerals.
- More stable minerals.

The major processes involved in chemical weathering are:

- Solution.
- Oxidation.
- Hydration and Hydrolysis.
- Carbonation – Dissolution.

Solution is the process of dissolving mineral constituents by water or acid. Most minerals have low solubility in pure water, but rain contains carbonic acid, so that carbonate minerals dissolve readily in acidic solutions.

Some substances present in the rocks are directly soluble in water. The soluble substances are removed by the continuous action of water and the rock no longer remains solid and form holes, rills or rough surface and ultimately falls into pieces or decomposes. The action is considerably increased when the water is acidified by the dissolution of organic and inorganic acids.

Hydration

Hydration involves two processes as hydration and hydrolysis. Hydration involves absorption of water. Chemical combination of water molecules with a particular substance or mineral leading to a change in structure. Soil forming minerals in rocks do not contain any water and they undergo hydration when exposed to humid conditions. Rain, streams, and seawater dissolve minerals from rocks, causing the rocks to crumble. For example water dissolves the mineral feldspar from granite, leaving grains of quartz, a mineral that forms sand.

Up on hydration there is swelling and increase in volume of minerals. The minerals lose their luster and become soft. It is one of the most common processes in nature and works with secondary minerals, such as aluminium oxide and iron oxide minerals and gypsum. Some of the typical mineral weathering reaction can easily illustrate these.

- Conversion of haematite into Limonite is a typical example.

$$2Fe_2O_3 + 3HOH \rightarrow 2Fe_2O_3 .3H_2O$$
$$(\text{Hematite})(\text{Red}) \quad (\text{Limonite})(\text{Yellow})$$

- Conversion of bauxite into hydrous aluminum oxide is another example.

$$Al_2O_3 + 3HOH \rightarrow Al_2O_3 .3H_2O$$
$$(\text{Bauxite}) \qquad (\text{Hyd. aluminium Oxide})$$

- Conversion of anhydrite into gypsum is another example.

$$CaSO_4 + 2H_2O \rightarrow CaSO_4 .2H_2O$$
$$(\text{Anhydrite}) \qquad (\text{Gypsum})$$

- Conversion of olivine into serpentine in yet another example.

$$3(MgO.FeO.SiO_2) + 2H_2O \rightarrow 3MgO.2SiO_2.2H_2O + SiO_2 + 3H_2O$$
$$(\text{Olivine}) \qquad (\text{Serpentine})$$

Hydrolysis is also a process of chemical weathering. It is due to the dissociation of H2O into H+ and OH- ions which chemically combine with minerals and bring about changes, such as exchange, decomposition of crystalline structure and formation of new compounds.

Water acts as a weak acid on silicate minerals. Reaction between mineral and water leads to the formation of a new mineral or dissolved material. Hydrolysis is the most important process in chemical weathering. It involves formation of hydroxyl ions and does not represent a chemical change. Hydrolysis of feldspar, for example, produces clay.

$$\underset{\text{(Orthoclase)}}{KAlSi_3O_8} + H_2O \rightarrow \underset{\text{(Aluminosilicic Acid + clay)}}{HAlSi_3O_8 + KOH}$$

Oxidation

Atmosphere contains 21% of free oxygen. Whereas the oxygen content of air dissolved in water is 30-35%. These two forms of oxygen are the most active chemical agents for weathering. Oxidation of minerals and rocks is an effective process in decoloration and decomposition of materials. In this process, the atmospheric oxygen combines with the metal ions of minerals to form oxides (or hydroxides). Sulphide minerals become unstable and gradually substituted by sulphates.

Oxidation (Attack by Oxygen)

$$2Fe_2SiO_4\,(\text{olivine}) + H_2O + O_2 = FeO.OH\,(\text{goethite}) + \text{dissolved silica}$$

Geothite dehydrates to form hematite, a very stable iron oxide. Geothite has a yellowish color, while hematite is brick red. Thus, rocks rich in iron oxides tend to form mainly red soils.

$$FeS_2 + 8HCO_2 + 7.5\,O_2 = Fe_2O_3 + 4SO_4^{-2} + 4CO_2 + 4H_2O$$

Pyrite oxidizes rapidly to form hematite in the presence of water and oxygen. The oxidation is more active in the presence of moisture and results in hydrated oxides. E.g., minerals containing Fe and Mg.

$$4FeO\,(\text{Ferrous oxide}) + O_2 \rightarrow 2Fe_2O_3\,(\text{Ferric oxide})$$
$$4Fe_3O_4\,(\text{Magnetite}) + O_2 \rightarrow 6Fe_2O_3\,(\text{Hematite})$$
$$2Fe_2O_3\,(\text{Hematite}) + 3H_2O \rightarrow 2Fe_2O_3\,.3H_2O\,(\text{Limonite})$$

The Minerals which are prone to chemical attack are Feldspars, Pyroxenes and Amphiboles. Minerals resistant to chemical attack are Quartz and Clay minerals (e.g Kaolinite).

Carbonation and Dissolution

Carbonation is the reaction of carbonate or carbonate ions with minerals. As rainwater falls through the atmosphere it dissolves small amounts of carbon dioxide. The atmospheric carbon dioxide when dissolved in water it forms carbonic acid. Carbonic acid plays a significant role in chemical weathering. Carbonic acid is formed naturally in rainwater. Additional carbon dioxide is picked-up in the ground from decaying vegetation. Carbonic acid is the principal weak acid- responsible for chemical weathering.

$$2H_2O + CO_2 \rightarrow H_2CO_3$$

This carbonic acid attacks many rocks and minerals and brings them into solution. The carbonated water has an etching effect up on some rocks, especially lime stone. The removal of cement that holds sand particles together leads to their disintegration.

$$CaCO_3 + H_2CO_3 \rightarrow Ca(HCO_3)_2$$

(Calcite) (Ca bicarbonate)

slightly soluble readily soluble

Acidic rainwater is also very effective at breaking down calcium carbonate.

Calcium carbonate + carbonic acid → calcium ions + bicarbonate ions

$$CaCO_3 + H_2CO_3 \rightarrow Ca^{++} + 2HCO_3^{-1}$$

Biological Weathering Processes

Unlike physical and chemical weathering, the biological or living agents are responsible for both decomposition and disintegration of rocks and minerals. The biological life is mainly controlled by the prevailing environment with its biotic and abiotic factors. Three aspects are to be understood in this as:

- Man and animals.
- Higher plants and roots.
- Micro-organisms.

Man and Animals

- The action of man in disintegration of rocks is well known as he cuts rocks to build dams, channels and construct roads and buildings. All these activities result in increasing the surface area of the rocks for attack of chemical agents and accelerate the process of decomposition.
- A large number of animals, birds, insects and worms, by their activities they make holes in them and thus aid for weathering.
- In tropical and sub-tropical regions, ants and termites build galleries and passages and carry materials from lower to upper surface and excrete acids.

- Rabbits, by burrowing in to the ground, destroy soft rocks. Moles, ants and bodies of the dead animals, provide substances which react with minerals and aid in decaying process.
- The earthworms pass the soil, through the alimentary canal and thus bring about physical and chemical changes in soil material.

Higher Plants and Roots

The roots of trees and other plants penetrate into the joints and crevices of the rocks. As they grew, they exert a great disruptive force and the hard rock may break apart. E.g., pipal tree growing on walls/rocks. Some roots penetrate deep into the soil and may open some sort of drainage channel. The roots running in crevices in lime stone and marble produces acids. These acids have a solvent action on carbonates.

The dead roots and plant residues decompose and produce carbon dioxide which is of great importance in weathering.

Micro-organisms

In early stages of mineral decomposition and soil formation, the lower forms of plants and animals like, mosses, bacteria and fungi and actinomycetes play an important role. They extract nutrients from the rock and N from air and live with a small quantity of water. In due course of time, the soil develops under the cluster of these micro-organisms. This organism closely associated with the decay of plant and animal remains and thus liberates nutrients for the use of next generation plants and also produces CO_2 and organic compounds which aid in mineral decomposition. Organisms can assist in breaking down the rocks into sediments or soils. Lichens, fungi and other micro-organisms are the typical examples.

Rate of Weathering

There are several factors determining the rate of weathering. The notable ones are:

- Climate.
- Biota (Organisms).
- Time.
- Mineral Composition.

Climate, which is usually measured in terms of temperature and moisture, can drastically affect the rate of weathering. High amounts of water and higher temperatures generally cause chemical reactions to run faster. Thus warm humid climates generally have more highly weathered rock, and rates of weathering are higher than in cold dry climates. Example: limestones in a dry desert climate are very resistant to weathering, but limestones in tropical climate weather very rapidly. A high temperature and high amounts of water also controls vegetation which indirectly affects the rate. Seasonality of precipitation affects the rate of weathering as well.

Organisms (Biota)

Animals-burrowing organisms like rodents, earthworms, and ants, bring material to the surface were it can be exposed to the agents of weathering.

Time

Weathering is a slow process. When we speak of time in a geologic sense, millions of years are often used as the units. The bedrock of Greenland has barely been weathered, and its age is over 3.5 billion years. The longer that a rock has been exposed to the weather, the faster it weathers.

Mineral Composition

All chemical and physical properties are determined by the mineral composition of the rock. Mineral composition is more important in determining the rate of chemical weathering.

Those minerals that are most reactive with acids, water, and air are weathered at a more rapid rate. Mineral composition also affects physical weathering. Minerals that are soft (hardness lower than six) can be easily abraded and blasted. Slope is yet another property for weathering: On steep slopes weathering products may be quickly washed away by rains. On gentle slopes the weathering products accumulate. On gentle slopes water may stay in contact with rock for longer periods of time, and thus result in higher weathering rates.

Porosity, faults and shears in materials are also controlling factors for weathering. Porous sediment equals better circulation and thus faster weathering process. Impermeable rocks will mean that no or little circulation will take place in rocks and weathering will be slower. Enhanced weathering rate occurs if faults and shears are present in rocks.

Exposure

Rock that is exposed to the atmosphere tends to weather much quicker. Bed rock that is covered by soil and vegetation may not weather as quickly.

Particle Size

The particle size of the material can affect the rate of weathering. The smaller the particle size, the faster it will weather. The more surface area that is exposed, the more it will weather.

Effect of Climate

Rapid weathering occurs in hot and wet climates. Cold, moderately dry climates experience intense mechanical weathering due to frost wedging. Cold, dry climates have very slow rates of rock weathering.

Behavior of Geologic Materials

Geologic materials are used in construction of buildings. These are subjected to weathering depending upon the rock's fundamental properties of hardness, durability, and stability over time. Mineral weathering plays a prominent role in many biogeochemical and geomorphological processes. Soil is the ultimate product of weathering. It supplies nutrients to soils and streams, accelerates physical erosion by weakening bedrock and producing easily erodible soil, and modulates Earth's long-term climate by drawing down atmospheric carbon dioxide. The temperature and rainfall of a region controls the conditions of weathering. When the temperature is very low mechanical weathering take the lead role. When the temperature increases and the rainfall is also more, chemical processes take the lead role.

Weathering is one of the major catalyst for promoting the rock cycle. It plays a very unique role in the rock cycle. Creation of soils and sediments are done by the weathering processes. Geomorphology plays a very important role on shaping the landscapes and landforms. From agriculture to human settlement, road networks to rail networks, industrial activities to all other development works, we depend on the landscapes and their properties. It is very essential to study these earth's dynamic processes, because most of our human activities are on these landscapes and landforms.

Physical Weathering

Physical weathering, also known as mechanical weather, is the process of rocks and minerals on Earth's surface breaking down or dissolving as a result of water, ice, salt, plants, animals or changes in temperature. Physical weathering does not change the chemical composition of the rock, just cracks and crumbles it into smaller pieces. After a rock has weathered, erosion occurs, transporting bits and pieces away. Finally a deposition process deposits the rock particles in a new place.

Weathering from Water

Water can weather rocks in a variety of ways. Moving water can lift and carry rocks from the bottom of a river or stream. When the rocks return to the ground under water, they can hit other rocks and break apart. Water can also weather a rock by affecting the material around it. For instance, clay that surrounds a rock can absorb water, swell and then push against the rock, causing it to break. Saltwater can cause another kind of weathering after it evaporates. When saltwater seeps into rock pore and then evaporate, crystals are left behind. The crystals grow and put pressure on the rock, which eventually causes it to break apart. Saltwater weathering is common along coastlines.

Weathering from Ice

When water sinks into cracks in a rock and the temperature drops low enough, the water freezes into ice. The ice expands and forms wedges in the rock that can split the rock into smaller fragments. Ice wedging usually occurs after water repeatedly freezes and melts inside small rock crevices over time. You can see the result of this type of weathering on street sidewalks in the winter. Ice wedges often cause potholes in roads and streets. Ice forms in the cracks of streets, expands and pushes on the surrounding rock or pavement, widening the cracks until they split and break apart.

Weathering from Plants

Plants can cause physical weathering as their roots grow. Seeds of plants or trees can grow inside rock cracks where soil has collected. The roots then put pressure on the cracks, making them wider and eventually splitting the rock. Even small plants can cause this kind of weathering over time.

Weathering from Animals

Animals that burrow underground, such as moles, gophers or even ants, can also cause physical weathering by loosening and breaking apart rocks. Dens and tunnels are signs of this type of weathering. Other animals dig and trample rock on the Earth's surface, causing rock to slowly crumble apart. This process exposes new parts of the rock to the elements, making them susceptible to other types of weathering, such as chemical weathering.

Chemical Weathering

Chemical weathering results from chemical changes to minerals that become unstable when they are exposed to surface conditions. The kinds of changes that take place are highly specific to the mineral and the environmental conditions. Some minerals, like quartz, are virtually unaffected by chemical weathering, while others, like feldspar, are easily altered. In general, the degree of chemical weathering is greatest in warm and wet climates, and least in cold and dry climates. The important characteristics of surface conditions that lead to chemical weathering are the presence of water (in the air and on the ground surface), the abundance of oxygen, and the presence of carbon dioxide, which produces weak carbonic acid when combined with water. That process, which is fundamental to most chemical weathering, can be shown as follows:

$$H_2O + CO_2 \rightarrow H_2CO_3 \text{ then } H_2CO_3 \rightarrow H^+ + HCO_3^-$$

Water + carbon dioxide → carbonic acid then carbonic acid → hydrogen ion + carbonate ion

Here we have water (e.g., as rain) plus carbon dioxide in the atmosphere, combining to create carbonic acid. Then carbonic acid dissociates (comes apart) to form hydrogen and carbonate ions. The amount of CO_2 in the air is enough to make only very weak carbonic acid, but there is typically much more CO_2 in the soil, so water that percolates through the soil can become significantly more acidic.

There are two main types of chemical weathering. On the one hand, some minerals become altered to other minerals. For example, feldspar is altered — by hydrolysis — to clay minerals. On the other hand, some minerals dissolve completely, and their components go into solution. For example, calcite ($CaCO_3$) is soluble in acidic solutions.

The hydrolysis of feldspar can be written like this:

$$CaAl_2Si_2O_8 + H_2CO_3 + ½O_2 \rightarrow Al_2Si_2O_5(OH)_4 + Ca^{2+} + CO_3^{2-}$$

Plagioclase + carbonic acid → kaolinite + dissolved calcium + carbonate ions

This reaction shows calcium plagioclase feldspar, but similar reactions could also be written for sodium or potassium feldspars. In this case, we end up with the mineral kaolinite, along with calcium and carbonate ions in solution. Those ions can eventually combine (probably in the ocean) to form the mineral calcite. The hydrolysis of feldspar to clay is illustrated in figure, which shows two images of the same granitic rock, a recently broken fresh surface on the left and a clay-altered weathered surface on the right. Other silicate minerals can also go through hydrolysis, although the end results will be a little different. For example, pyroxene can be converted to the clay minerals chlorite or smectite, and olivine can be converted to the clay mineral serpentine.

Unweathered (left) and weathered (right) surfaces of the same piece of granitic rock. On the unweathered surfaces the feldspars are still fresh and glassy-looking. On the weathered surface the feldspar has been altered to the chalky-looking clay mineral kaolinite.

Oxidation is another very important chemical weathering process. The oxidation of the iron in a ferromagnesian silicate starts with the dissolution of the iron. For olivine, the process looks like this, where olivine in the presence of carbonic acid is converted to dissolved iron, carbonate, and silicic acid:

$$Fe_2SiO_4 + 4H_2CO_3 \rightarrow 2Fe^{2+} + 4HCO_3^- + H_4SiO_4$$

Olivine + (carbonic acid) → dissolved iron + dissolved carbonate + dissolved silicic acid

In the presence of oxygen, the dissolved iron is then quickly converted to hematite:

$$2Fe^{2+} + 4HCO_3^- + ½O_2 + 2H_2O \rightarrow Fe_2O_3 + 4H_2CO_3$$

dissolved iron + bicarbonate + oxygen + water →hematite + carbonic acid

The equation shown here is for olivine, but it could apply to almost any other ferromagnesian silicate, including pyroxene, amphibole, or biotite. Iron in the sulphide minerals (e.g., pyrite) can also be oxidized in this way. And the mineral hematite is not the only possible end result, as there is a wide range of iron oxide minerals that can form in this way. The results of this process are illustrated in figure, which shows a granitic rock in which some of the biotite and amphibole have been altered to form the iron oxide mineral limonite.

A granitic rock containing biotite and amphibole which have been altered near to the rock's surface to limonite, which is a mixture of iron oxide minerals.

A special type of oxidation takes place in areas where the rocks have elevated levels of sulphide minerals, especially pyrite (FeS_2). Pyrite reacts with water and oxygen to form sulphuric acid, as follows:

$$2FeS_2 + 7O_2 + 2H_2O \rightarrow 2Fe^{2+} \; H_2SO_4 + 2H^+$$

Pyrite + oxygen + water → iron ions + sulphuric acid + hydrogen ions

The runoff from areas where this process is taking place is known as acid rock drainage (ARD), and even a rock with 1% or 2% pyrite can produce significant ARD. Some of the worst examples of ARD are at metal mine sites, especially where pyrite-bearing rock and waste material have been mined from deep underground and then piled up and left exposed to water and oxygen. One example of that is the Mt. Washington Mine near Courtenay on Vancouver Island, but there are many similar sites across Canada and around the world.

Exposed oxidizing and acid generating rocks and mine waste at the abandoned Mt. Washington Mine, B.C. (left), and an example of acid drainage downstream from the mine site (right).

At many ARD sites, the pH of the runoff water is less than 4 (very acidic). Under these conditions, metals such as copper, zinc, and lead are quite soluble, which can lead to toxicity for aquatic and other organisms. For many years, the river downstream from the Mt. Washington Mine had so much dissolved copper in it that it was toxic to salmon. Remediation work has since been carried out at the mine and the situation has improved.

The hydrolysis of feldspar and other silicate minerals and the oxidation of iron in ferromagnesian silicates all serve to create rocks that are softer and weaker than they were to begin with, and thus more susceptible to mechanical weathering.

The weathering reactions involved the transformation of one mineral to another mineral (e.g., feldspar to clay), and the release of some ions in solution (e.g., Ca^{2+}). Some weathering processes involve the complete dissolution of a mineral. Calcite, for example, will dissolve in weak acid, to produce calcium and bicarbonate ions. The equation is as follows:

$$CaCO_3 + H^+ + HCO_3^- \rightarrow Ca^{2+} + 2HCO_3^- .$$

Calcite + Hydrogen ions + Bicarbonate → Calcium Ions + Bicarbonate

Calcite is the major component of limestone (typically more than 95%), and under surface conditions, limestone will dissolve to varying degrees (depending on which minerals it contains, other than calcite), as shown in figure. Limestone also dissolves at relatively shallow depths underground, forming limestone caves.

Limestone outcrop on Quadra Island, B.C. The limestone, which is primarily made up of the mineral calcite, has been dissolved to different degrees in different areas because of compositional differences. The buff-coloured bands are volcanic rock, which is not soluble.

Types and Examples of Chemical Weathering

Water, acids, and oxygen are just a few of the chemicals that lead to geological change. Over time, chemical weathering can produce dramatic results.

Chemical Weathering from Water

Water causes both mechanical weathering and chemical weathering. Mechanical weathering occurs when water drips or flows over rock for prolonged periods; the Grand Canyon, for example, was formed to a large degree by the mechanical weathering action of the Colorado River.

Chemical weathering occurs when water dissolves minerals in a rock, producing new compounds. This reaction is called hydrolysis. Hydrolysis occurs, for example, when water comes in contact with granite. Feldspar crystals inside the granite react chemically, forming clay minerals. The clay weakens the rock, making it more likely to break.

Water also interacts with calcites in caves, causing them to dissolve. Calcite in dripping water builds up over many years to create stalagmites and stalactites.

In addition to changing the shapes of rocks, chemical weathering from water changes the composition of water. For example, weathering over billions of years is a big factor in why the ocean is salty.

Chemical Weathering from Oxygen

Oxygen is a reactive element. It reacts with rocks through a process called oxidation. One example of this type of weathering is rust formation, which occurs when oxygen reacts with iron to form iron oxide (rust). Rust changes the color of the rocks, plus iron oxide is much more fragile than iron, so the weathered region becomes more susceptible to breakage.

Chemical Weathering from Acids

When rocks and minerals are altered by hydrolysis, acids may be produced. Acids may also be produced when water reacts with the atmosphere, so acidic water can react with rocks. The effect of acids on minerals is an example of solution weathering. Solution weathering also covers other types of chemical solutions, such as basic rather than acidic ones.

One common acid is carbonic acid, a weak acid that is produced when carbon dioxide reacts with water. Carbonation is an important process in the formation of many caves and sinkholes. Calcite in limestone dissolves under acidic conditions, leaving open spaces.

Chemical Weathering from Living Organisms

Living organisms perform chemical reactions to obtain minerals from soil and rocks. Many chemical changes are possible.

Lichens can have a profound effect on rock. Lichens, a combination of algae and fungi, produce a weak acid that can dissolve rock. Plant roots are also an important source of chemical weathering. As roots expand into rock, acids can change the minerals in the rock. Plant roots also use carbon dioxide, thus changing the chemistry of the soil.

New, weaker minerals are often more brittle; this makes it easier for plant roots to break up the rock. Once the rock is broken up, water can get into the cracks and oxidize or freeze. Frozen water expands, making the cracks wider and further weathering the rock.

Animals can also affect geochemistry. For example, bat guano and other animal remains contain reactive chemicals that can affect minerals.

Human activities also have a major impact on rock. Mining, of course, changes the location and condition of rocks and soil. Acid rain caused by pollution can eat away at rocks and minerals. Farming changes the chemical composition of soil, mud, and rock.

Spheroidal Weathering

Spheroidal or woolsack weathering in granite on Haytor, Dartmoor.

Spheroidal weathering is a form of chemical weathering that affects jointed bedrock and results in the formation of concentric or spherical layers of highly decayed rock within weathered bedrock that is known as saprolite. When saprolite is exposed by physical erosion, these concentric layers peel (spall) off as concentric shells much like the layers of a peeled onion. Within saprolite, spheroidal weathering often creates rounded boulders, known as corestones or woolsack, of relatively unweathered rock. Spheroidal weathering is also called onion skin weathering, concentric weathering, spherical weathering, or woolsack weathering.

Spheroidal weathering in granite, Estaca de Bares, A Coruña, Galicia.

Woolsack weathering in sandstone at the Externsteine rocks, Teutoburg Forest.

Corestones near Musina that were created by spherodial weathering and exposed by the removal of surrounding saprolite by erosion.

Spheroidal weathering of a dolerite dyke, Pilbara.

Weathering Process

Spheroidal weathering is the result of chemical weathering of systematically jointed, massive rocks, including granite, dolerite, basalt and sedimentary rocks such as silicified sandstone. It occurs as the result of the chemical alteration of such rocks along intersecting joints. The chemical alteration of the rock results in the formation of abundant secondary minerals such as kaolinite, sericite, serpentine, montmorillonite, and chlorite and a corresponding increase in the volume of the altered rock. When the joints within bedrock form a 3-dimensional network, they subdivided it into separated blocks, often in the form of cubes or rectangles, that are bounded by these joints. Because water can penetrate the bedrock along these joints, the near-surface bedrock will be altered by weathering progressively inward along the faces of these blocks. The alteration by weathering of the bedrock will be greatest along the corners of each block, followed by the edges, and finally the faces of the cube. The differences in weathering rates between the corners, edges, and faces of a bedrock block will result in the formation of spheroidal layers of altered rock that surround an unaltered rounded boulder-size core of relatively unaltered rock known as a corestone or woolsack. Spheroidal weathering has often been incorrectly attributed solely to various types of physical weathering.

Frequently, erosion has removed the layers of altered rock and other saprolite surrounding corestones that were produced by spheroidal weathering. This leaves many corestones as freestanding boulders on the ground's surface. Often the spheroidal weathering, which created these corestones and the enclosing saprolite occurred in the prehistoric past during periods of humid, even tropical climates. Frequently, the removal of the saprolite by erosion and exposure of corestones as freestanding residual boulders, tors, or other landforms occurs many thousands of years later and during vastly different climatic conditions.

Depending on local environmental conditions, spheroidal weathering of bedrock blocks defined by tectonically induced joints and fractures may result in the formation of prominent and well-defined Liesegang rings within these blocks. These blocks typically consist of bedrock blocks (Liesegang blocks), which are bounded on their periphery by joints and fractures, and, in sedimentary rocks, bedding planes above and below. Each Liesegang block consists of a relatively unaltered core surrounded by concentric, alternating shells of iron-poor (intermediate shells) and iron-rich ('iron' shells) composition which make up the Liesegang rings. These iron-poor and iron-rich shells follow the configuration of the outer shape of the block and are sub-parallel to its sides. The iron-rich and iron-poor shells vary in degree of cementation and, as a result, can produce box work weathering structures during subsequent erosion. The degree of development of Liesegang rings as the result of weathering depends upon the spacing of the joint systems, groundwater flow, local topography, bedrock composition, and bed thickness.

Honeycomb Weathering

Honeycomb weathering, also known as honeycombs, honeycombed sandstone, is a form of cavernous weathering and subcategory of tafoni that consists of regular, tightly adjoining, and commonly patterned cavities that are developed in weathered bedrock; are less than 2 cm (0.79 in) in size; and resemble a honeycombed structure. Honeycombs also been called alveoli, lacework, stonelace, fretting or miniature tafoni weathering. The size at which honeycombs are differentiated from tafoni varies greatly in the scientific literature and lacks an official consensus.

Honeycomb weathering in Larrabee State Park, Washington.

Honeycomb weathering in a Cambrian sandstone, Timna Valley, Negev Desert, Israel.

Honeycomb weathering at Altdahn Castle in the Palatinate Forest, Germany.

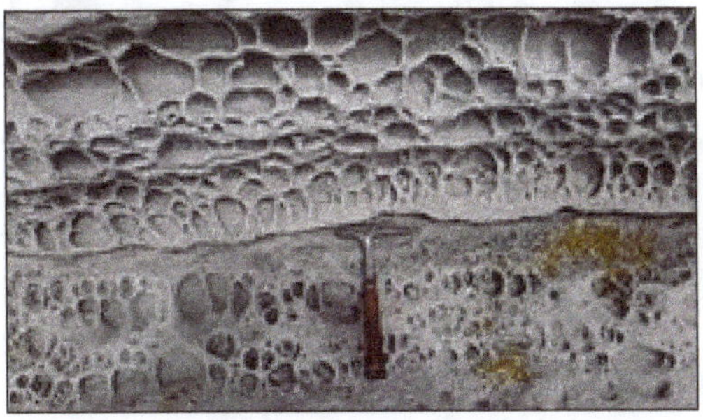

Excellent honeycomb weathering of a cliff overhang, Elgol, Scotland.

Distribution

Honeycomb weathering typically develops in siliceous, either coarse-grained sedimentary (sandstone) or coarsely crystalline plutonic (granite) rocks. It can be found in all climate types, but is most prolific in salt-rich environments, such as deserts and coastal zones. The common factors in the environments in which it is found are high salt concentrations and frequent or occasional desiccating conditions. Honeycomb weathering is rarely associated with highly soluble rock such as limestone or dolomite. Honeycombs also occur in stone structures, e.g. buildings and breakwaters, where a rate of development can be established. Honeycomb weathering has been found to have formed in greywacke blocks since they were used to build seawalls on the coast of Southeast Australia in 1943 and 1949.

Honeycomb and other forms of cavernous weathering are likely as characteristic of the weathering of sandstones as fluting is of limestones, but it is not uniformly or universally developed. For example, in some cliffs, large expanses of the rock face exhibit surface weathering and caverous weathering of metric dimensions. However, other rock faces are pitted between closely spaced bedding and vertical joint planes. In the southwestern United States, sheer faces of Coconino and Supai Sandstones cropping out along the sides of the Grand Canyon are only sparsely honeycombed. Also, the tall spires of De Chelley Sandstone at Monument Valley in Utah are unmarked by such weathering. In contrast, the Aztec Sandstone of The Valley of Fire in Nevada exhibits delicate and widespread honeycomb and other cavernous weathering. The difference may lie in contrasting patterns of pearmeability and jointing and resulting variations in the flow of waters through these sandstone.

Cause

Many explanations have been proposed for honeycomb and other cavernous weathering. These explanations include marine abrasion; wind corrosion; mechanical weathering resulting from short-term temperature variations; chemical weathering of the interior of the rock (core-softening) under a protective crust (case-hardening) followed by mechanical removal of the softened material; biogeochemical weathering by lichens; temperature variations acting on salt efflorescence in coastal regions; and salt weathering. Most commonly, researchers have advocated salt weathering as the primary explanation for the formation of honeycomb weathering. Currently, it is considered to be polygenetic in origin; being the result of complex interaction of physical and chemical weathering processes, which include salt weathering and cyclic wetting and drying. There are instances where honeycombing or pitting of sandstone is due simply to removal of easily soluble cements such as calcite or dolomite, e.g. Cretaceous sandstones in central Kansas.

Frost Weathering

A rock in Abisko, Sweden fractured (along existing joints) possibly by mechanical frost weathering or thermal stress; a chullo is shown for scale.

Frost weathering is a collective term for several mechanical weathering processes induced by stresses created by the freezing of water into ice. The term serves as an umbrella term for a variety of processes such as frost shattering, frost wedging and cryofracturing. The process may act on a wide range of spatial and temporal scales, from minutes to years and from dislodging mineral grains to fracturing boulders. It is most pronounced in high-altitude and high-latitude areas and is especially associated with alpine, periglacial, subpolar maritime and polar climates, but may occur anywhere at sub-freezing temperatures (between -3 and -8°C) if water is present.

Ice Segregation

Certain frost-susceptible soils expand or heave upon freezing as a result of water migrating via capillary action to grow ice lenses near the freezing front. This same phenomenon occurs within pore spaces of rocks. The ice accumulations grow larger as they attract liquid water from the surrounding pores. The ice crystal growth weakens the rocks which, in time, break up. It is caused by the expansion of ice when water freezes, putting considerable stress on the walls of containment. This is actually a very common process in all humid, temperate areas where there is exposed rock, especially porous rocks like sandstone. Sand can often be found just under the faces of exposed sandstone where individual grains have been popped off, one by one. This process is often termed frost spalling. In fact, this is often the most important weathering process for exposed rock in many areas.

Similar processes can act on asphalt pavements, contributing to various forms of cracking and other distresses, which, when combined with traffic and the intrusion of water, accelerate rutting, the formation of potholes, and other forms of pavement roughness.

Volumetric Expansion

The traditional explanation for frost weathering was volumetric expansion of freezing water. When water freezes to ice, its volume increases by nine percent. Under specific circumstances, this expansion is able to displace or fracture rock. At a temperature of -22°C, ice growth is known to be able to generate pressures of up to 207MPa, more than enough to fracture any rock. For frost weathering to occur by volumetric expansion, the rock must have almost no air that can be compressed to compensate for the expansion of ice, which means it has to be water-saturated and frozen quickly from all sides so that the water does not migrate away and the pressure is exerted on the rock. These conditions are considered unusual, restricting it to a process of importance within a few centimeters of a rock's surface and on larger existing water-filled joints in a process called ice wedging.

Not all volumetric expansion is caused by the pressure of the freezing water; it can be caused by stresses in water that remains unfrozen. When ice growth induces stresses in the pore water that breaks the rock, the result is called hydrofracture. Hydrofracturing is favoured by large interconnected pores or large hydraulic gradients in the rock. If there are small pores, a very quick freezing of water in parts of the rock may expel water, and if the water is expelled faster than it can migrate, pressure may rise, fracturing the rock.

Since research in physical weathering begun around 1900, volumetric expansion was, until the 1980s, held to be the predominant process behind frost weathering. This view was challenged in 1985 and 1986 publications by Walder and Hallet. Nowadays researchers such as Matsuoka and Murton consider the "conditions necessary for frost weathering by volumetric expansion" as unusual. However the bulk of recent literature demonstrates that that ice segregation is capable of providing quantitative models for common phenomena while the traditional, simplistic volumetric expansion does not.

Erosion

An actively eroding rill on an intensively-farmed field in eastern Germany.

A natural arch produced by the wind erosion of differentially weathered rock in Jebel Kharaz, Jordan.

In earth science, erosion is the action of surface processes (such as water flow or wind) that removes soil, rock, or dissolved material from one location on the Earth's crust, and then transports it to another location. This natural process is caused by the dynamic activity of erosive agents, that is, water, ice (glaciers), snow, air (wind), plants, animals, and humans. In accordance with these agents, erosion is sometimes divided into water erosion, glacial erosion, snow erosion, wind (aeolic) erosion, zoogenic erosion, and anthropogenic erosion. The particulate breakdown of rock or soil into clastic sediment is referred to as physical or mechanical erosion; this contrasts with chemical erosion, where soil or rock material is removed from an area by its dissolving into a solvent (typically water), followed by the flow away of that solution. Eroded sediment or solutes may be transported just a few millimetres, or for thousands of kilometres.

Natural rates of erosion are controlled by the action of geological weathering geomorphic drivers, such as rainfall; bedrock wear in rivers; coastal erosion by the sea and waves; glacial plucking, abrasion, and scour; areal flooding; wind abrasion; groundwater processes; and mass movement processes in steep landscapes like landslides and debris flows. The rates at which such processes act control how fast a surface is eroded. Typically, physical erosion proceeds fastest on steeply sloping surfaces, and rates may also be sensitive to some climatically-controlled properties including amounts of water supplied (e.g., by rain), storminess, wind speed, wave fetch, or atmospheric temperature (especially for some ice-related processes). Feedbacks are also possible between rates of erosion and the amount of eroded material that is already carried by, for example, a river or glacier. Processes of erosion that produce sediment or solutes from a place contrast with those of deposition, which control the arrival and emplacement of material at a new location.

A wave-like sea cliff produced by coastal erosion, in Jinshitan Coastal National Geopark, Dalian, Liaoning Province, China.

While erosion is a natural process, human activities have increased by 10-40 times the rate at which erosion is occurring globally. At well-known agriculture sites such as the Appalachian Mountains, intensive farming practices have caused erosion up to 100x the speed of the natural rate of erosion in the region. Excessive (or accelerated) erosion causes both "on-site" and "off-site" problems. On-site impacts include decreases in agricultural productivity and (on natural landscapes) ecological collapse, both because of loss of the nutrient-rich upper soil layers. In some cases, the eventual end result is desertification. Off-site effects include sedimentation of waterways and eutrophication of water bodies, as well as sediment-related damage to roads and houses. Water and wind erosion are the two primary causes of land degradation; combined, they are responsible for about 84% of the global extent of degraded land, making excessive erosion one of the most significant environmental problems worldwide.

Intensive agriculture, deforestation, roads, anthropogenic climate change and urban sprawl are amongst the most significant human activities in regard to their effect on stimulating erosion. However, there are many prevention and remediation practices that can curtail or limit erosion of vulnerable soils.

Physical Processes

Rainfall and Surface Runoff

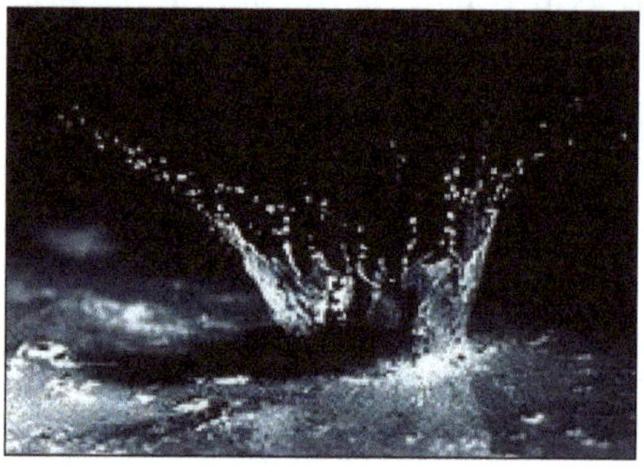

Soil and water being splashed by the impact of a single raindrop.

Rainfall, and the surface runoff which may result from rainfall, produces four main types of soil erosion: splash erosion, sheet erosion, rill erosion, and gully erosion. Splash erosion is generally seen as the first and least severe stage in the soil erosion process, which is followed by sheet erosion, then rill erosion and finally gully erosion (the most severe of the four).

In splash erosion, the impact of a falling raindrop creates a small crater in the soil, ejecting soil particles. The distance these soil particles travel can be as much as 0.6 m (two feet) vertically and 1.5 m (five feet) horizontally on level ground.

If the soil is saturated, or if the rainfall rate is greater than the rate at which water can infiltrate into the soil, surface runoff occurs. If the runoff has sufficient flow energy, it will transport loosened soil particles (sediment) down the slope. Sheet erosion is the transport of loosened soil particles by overland flow.

A spoil tip covered in rills and gullies due to erosion processes caused by rainfall: Rummu, Estonia.

Rill erosion refers to the development of small, ephemeral concentrated flow paths which function as both sediment source and sediment delivery systems for erosion on hillslopes. Generally, where water erosion rates on disturbed upland areas are greatest, rills are active. Flow depths in rills are typically of the order of a few centimetres (about an inch) or less and along-channel slopes may be quite steep. This means that rills exhibit hydraulic physics very different from water flowing through the deeper, wider channels of streams and rivers.

Gully erosion occurs when runoff water accumulates and rapidly flows in narrow channels during or immediately after heavy rains or melting snow, removing soil to a considerable depth.

Rivers and Streams

Dobbingstone Burn, Scotland, showing two different types of erosion affecting the same place. Valley erosion is occurring due to the flow of the stream, and the boulders and stones (and much of the soil) that are lying on the stream's banks are glacial till that was left behind as ice age glaciers flowed over the terrain.

Valley or stream erosion occurs with continued water flow along a linear feature. The erosion is both downward, deepening the valley, and headward, extending the valley into the hillside, creating head cuts and steep banks. In the earliest stage of stream erosion, the erosive activity is dominantly vertical, the valleys have a typical V cross-section and the stream gradient is relatively steep. When some base level is reached, the erosive activity switches to lateral erosion, which widens the valley floor and creates a narrow floodplain. The stream gradient becomes nearly flat, and lateral deposition of sediments becomes important as the stream meanders across the valley floor. In all stages of stream erosion, by far the most erosion occurs during times of flood when more and faster-moving water is available to carry a larger sediment load. In such processes, it is not the water alone that erodes: suspended abrasive particles, pebbles, and boulders can also act erosively as they traverse a surface, in a process known as traction.

Bank erosion is the wearing away of the banks of a stream or river. This is distinguished from changes on the bed of the watercourse, which is referred to as scour. Erosion and changes in the form of river banks may be measured by inserting metal rods into the bank and marking the position of the bank surface along the rods at different times.

Thermal erosion is the result of melting and weakening permafrost due to moving water. It can occur both along rivers and at the coast. Rapid river channel migration observed in the Lena River of Siberia is due to thermal erosion, as these portions of the banks are composed of permafrost-cemented non-cohesive materials. Much of this erosion occurs as the weakened banks fail in large slumps. Thermal erosion also affects the Arctic coast, where wave action and near-shore temperatures combine to undercut permafrost bluffs along the shoreline and cause them to fail. Annual erosion rates along a 100-kilometre (62-mile) segment of the Beaufort Sea shoreline averaged 5.6 metres (18 feet) per year from 1955 to 2002.

Most river erosion happens nearer to the mouth of a river. On a river bend, the longest least sharp side has slower moving water. Here deposits build up. On the narrowest sharpest side of the bend, there is faster moving water so this side tends to erode away mostly.

Coastal Erosion

Wave cut platform caused by erosion of cliffs by the sea, at Southerndown in South Wales.

Erosion of the boulder clay (of Pleistocene age) along cliffs of Filey Bay, Yorkshire, England.

Shoreline erosion, which occurs on both exposed and sheltered coasts, primarily occurs through the action of currents and waves but sea level (tidal) change can also play a role.

Hydraulic action takes place when the air in a joint is suddenly compressed by a wave closing the entrance of the joint. This then cracks it. Wave pounding is when the sheer energy of the wave hitting the cliff or rock breaks pieces off. Abrasion or corrasion is caused by waves launching sea load at the cliff. It is the most effective and rapid form of shoreline erosion. Corrosion is the dissolving of rock by carbonic acid in sea water. Limestone cliffs are particularly vulnerable to this kind of erosion. Attrition is where particles/sea load carried by the waves are worn down as they hit each other and the cliffs. This then makes the material easier to wash away. The material ends up as shingle and sand. Another significant source of erosion, particularly on carbonate coastlines, is boring, scraping and grinding of organisms, a process termed bioerosion.

Sediment is transported along the coast in the direction of the prevailing current (longshore drift). When the upcurrent amount of sediment is less than the amount being carried away, erosion occurs. When the upcurrent amount of sediment is greater, sand or gravel banks will tend to form as a result of deposition. These banks may slowly migrate along the coast in the direction of the longshore drift, alternately protecting and exposing parts of the coastline. Where there is a bend in the coastline, quite often a buildup of eroded material occurs forming a long narrow bank (a spit). Armoured beaches and submerged offshore sandbanks may also protect parts of a coastline from erosion. Over the years, as the shoals gradually shift, the erosion may be redirected to attack different parts of the shore.

Chemical Erosion

Chemical erosion is the loss of matter in a landscape in the form of solutes. Chemical erosion is usually calculated from the solutes found in streams. Anders Rapp pioneered the study of chemical erosion in his work about Kärkevagge published in 1960.

Glaciers

Glacial moraines above Lake Louise, in Alberta, Canada.

Glaciers erode predominantly by three different processes: abrasion/scouring, plucking, and ice thrusting. In an abrasion process, debris in the basal ice scrapes along the bed, polishing and gouging the underlying rocks, similar to sandpaper on wood. Scientists have shown that, in addition to the role of temperature played in valley-deepening, other glaciological processes, such as erosion also control cross-valley variations. In a homogeneous bedrock erosion pattern, curved channel cross-section beneath the ice is created. Though the glacier continues to incise vertically, the shape of the channel beneath the ice eventually remain constant, reaching a U-shaped parabolic steady-state shape as we now see in glaciated valleys. Scientists also provide a numerical estimate of the time required for the ultimate formation of a steady-shaped U-shaped valley—approximately 100,000 years. In a weak bedrock (containing material more erodible than the surrounding rocks) erosion pattern, on the contrary, the amount of over deepening is limited because ice velocities and erosion rates are reduced.

Glaciers can also cause pieces of bedrock to crack off in the process of plucking. In ice thrusting, the glacier freezes to its bed, and then as it surges forward, it moves large sheets of frozen sediment at the base along with the glacier. This method produced some of the many thousands of lake basins that dot the edge of the Canadian Shield. Differences in the height of mountain ranges are not only being the result tectonic forces, such as rock uplift, but also local climate variations. Scientists use global analysis of topography to show that glacial erosion controls the maximum height of mountains, as the relief between mountain peaks and the snow line are generally confined to altitudes less than 1500 m. The erosion caused by glaciers worldwide erodes mountains so effectively that the term glacial buzzsaw has become widely used, which describes the limiting effect of glaciers on the height of mountain ranges. As mountains grow higher, they generally allow for more glacial activity (especially in the accumulation zone above the glacial equilibrium line altitude), which causes increased rates of erosion of the mountain, decreasing mass faster than isostatic rebound can add to the mountain. This provides a good example of a negative feedback loop. Ongoing research is showing that while glaciers tend to decrease mountain size, in some areas, glaciers can actually reduce the rate of erosion, acting as a glacial armor. Ice can not only erode mountains but also protect them from erosion. Depending on glacier regime, even steep alpine lands can be preserved through time with the help of ice. Scientists have proved this theory by sampling eight summits of northwestern Svalbard using Be10 and Al26, showing that northwestern Svalbard transformed from a glacier-erosion state under relatively mild glacial maxima temperature, to a glacier-armor state occupied by cold-based, protective ice during much colder glacial maxima temperatures as the Quaternary ice age progressed.

These processes, combined with erosion and transport by the water network beneath the glacier, leave behind glacial landforms such as moraines, drumlins, ground moraine (till), kames, kame deltas, moulins, and glacial erratics in their wake, typically at the terminus or during glacier retreat.

The best-developed glacial valley morphology appears to be restricted to landscapes with low rock uplift rates (less than or equal to 2 mm per year) and high relief, leading to long-turnover times. Where rock uplift rates exceed 2 mm per year, glacial valley morphology has generally been significantly modified in postglacial time. Interplay of glacial erosion and tectonic forcing governs the morphologic impact of glaciations on active orogens, by both influencing their height, and by altering the patterns of erosion during subsequent glacial periods via a link between rock uplift and valley cross-sectional shape.

Floods

At extremely high flows, kolks, or vortices are formed by large volumes of rapidly rushing water. Kolks cause extreme local erosion, plucking bedrock and creating pothole-type geographical features called Rock-cut basins. Examples can be seen in the flood regions result from glacial Lake Missoula, which created the channeled scablands in the Columbia Basin region of eastern Washington.

Wind Erosion

Árbol de Piedra, a rock formation in the Altiplano, Bolivia sculpted by wind erosion.

Wind erosion is a major geomorphological force, especially in arid and semi-arid regions. It is also a major source of land degradation, evaporation, desertification, harmful airborne dust, and crop damage—especially after being increased far above natural rates by human activities such as deforestation, urbanization, and agriculture.

Wind erosion is of two primary varieties: deflation, where the wind picks up and carries away loose particles; and abrasion, where surfaces are worn down as they are struck by airborne particles carried by wind. Deflation is divided into three categories: (1) surface creep, where larger, heavier particles slide or roll along the ground; (2) saltation, where particles are lifted a short height into the air, and bounce and saltate across the surface of the soil; and (3) suspension, where very small and light particles are lifted into the air by the wind, and are often carried for long distances. Saltation is responsible for the majority (50-70%) of wind erosion, followed by suspension (30-40%), and then surface creep (5-25%).

Wind erosion is much more severe in arid areas and during times of drought. For example, in the Great Plains, it is estimated that soil loss due to wind erosion can be as much as 6100 times greater in drought years than in wet years.

Mass Movement

Mass movement is the downward and outward movement of rock and sediments on a sloped surface, mainly due to the force of gravity.

Mass movement is an important part of the erosional process and is often the first stage in the breakdown and transport of weathered materials in mountainous areas. It moves material from higher elevations to lower elevations where other eroding agents such as streams and glaciers can then pick up the material and move it to even lower elevations. Mass-movement processes are always occurring continuously on all slopes; some mass-movement processes act very slowly; others occur very suddenly, often with disastrous results. Any perceptible down-slope movement of rock or sediment is often referred to in general terms as a landslide. However, landslides can be classified in a much more detailed way that reflects the mechanisms responsible for the movement and the velocity at which the movement occurs. One of the visible topographical manifestations of a very slow form of such activity is a scree slope.

A wadi in Makhtesh Ramon, Israel, showing gravity collapse erosion on its banks.

Slumping happens on steep hillsides, occurring along distinct fracture zones, often within materials like clay that, once released, may move quite rapidly downhill. They will often show a spoon-shaped isostatic depression, in which the material has begun to slide downhill. In some cases, the slump is caused by water beneath the slope weakening it. In many cases it is simply the result of poor engineering along highways where it is a regular occurrence.

Surface creep is the slow movement of soil and rock debris by gravity which is usually not perceptible except through extended observation. However, the term can also describe the rolling of dislodged soil particles 0.5 to 1.0 mm (0.02 to 0.04 in) in diameter by wind along the soil surface.

Factors affecting Erosion Rates

Climate

The amount and intensity of precipitation is the main climatic factor governing soil erosion by water. The relationship is particularly strong if heavy rainfall occurs at times when, or in locations where, the soil's surface is not well protected by vegetation. This might be during periods when agricultural activities leave the soil bare, or in semi-arid regions where vegetation is naturally sparse. Wind erosion requires strong winds, particularly during times of drought when vegetation is sparse and soil is dry (and so is more erodible). Other climatic factors such as average temperature and temperature range may also affect erosion, via their effects on vegetation and soil properties. In general, given similar vegetation and ecosystems, areas with more precipitation (especially high-intensity rainfall), more wind, or more storms are expected to have more erosion.

In some areas of the world (e.g. the mid-western USA), rainfall intensity is the primary determinant of erosivity (for a definition of erosivity check,) with higher intensity rainfall generally resulting in more soil erosion by water. The size and velocity of rain drops is also an important factor. Larger and higher-velocity rain drops have greater kinetic energy, and thus their impact will displace soil particles by larger distances than smaller, slower-moving rain drops.

In other regions of the world (e.g. western Europe), runoff and erosion result from relatively low intensities of stratiform rainfall falling onto the previously saturated soil. In such situations, rainfall amount rather than intensity is the main factor determining the severity of soil erosion by water.

In Taiwan, where typhoon frequency increased significantly in the 21st century, a strong link has been drawn between the increase in storm frequency with an increase in sediment load in rivers and reservoirs, highlighting the impacts climate change can have on erosion.

Vegetative Cover

Vegetation acts as an interface between the atmosphere and the soil. It increases the permeability of the soil to rainwater, thus decreasing runoff. It shelters the soil from winds, which results in decreased wind erosion, as well as advantageous changes in microclimate. The roots of the plants bind the soil together, and interweave with other roots, forming a more solid mass that is less susceptible to both water and wind erosion. The removal of vegetation increases the rate of surface erosion.

Topography

The topography of the land determines the velocity at which surface runoff will flow, which in turn determines the erosivity of the runoff. Longer, steeper slopes (especially those without adequate vegetative cover) are more susceptible to very high rates of erosion during heavy rains than shorter, less steep slopes. Steeper terrain is also more prone to mudslides, landslides, and other forms of gravitational erosion processes.

Tectonics

Tectonic processes control rates and distributions of erosion at the Earth's surface. If the tectonic action causes part of the Earth's surface (e.g., a mountain range) to be raised or lowered relative to surrounding areas, this must necessarily change the gradient of the land surface. Because erosion rates are almost always sensitive to the local slope, this will change the rates of erosion in the uplifted area. Active tectonics also brings fresh, unweathered rock towards the surface, where it is exposed to the action of erosion.

However, erosion can also affect tectonic processes. The removal by erosion of large amounts of rock from a particular region, and its deposition elsewhere, can result in a lightening of the load on the lower crust and mantle. Because tectonic processes are driven by gradients in the stress field developed in the crust, this unloading can in turn cause tectonic or isostatic uplift in the region. In some cases, it has been hypothesised that these twin feedbacks can act to localize and enhance zones of very rapid exhumation of deep crustal rocks beneath places on the Earth's surface with extremely high erosion rates, for example, beneath the extremely steep terrain of Nanga Parbat in the western Himalayas. Such a place has been called a "tectonic aneurysm".

Development

Human land development, in forms including agricultural and urban development, is considered a significant factor in erosion and sediment transport. In Taiwan, increases in sediment load in the northern, central, and southern regions of the island can be tracked with the timeline of development for each region throughout the 20th century.

Erosion at Various Scales

Mountain Ranges

Mountain ranges are known to take many millions of years to erode to the degree they effectively cease to exist. Scholars Pitman and Golovchenko estimate that it takes probably more than 450 million years to erode a mountain mass similar to the Himalaya into an almost-flat peneplain if there are no major sea-level changes. Erosion of mountains massifs can create a pattern of equally high summits called summit accordance. It has been argued that extension during post-orogenic collapse is a more effective mechanism of lowering the height of orogenic mountains than erosion.

Examples of heavily eroded mountain ranges include the Timanides of Northern Russia. Erosion of this orogen has produced sediments that are now found in the East European Platform, including the Cambrian Sablya Formation near Lake Ladoga. Studies of these sediments indicate that it is likely that the erosion of the orogen began in the Cambrian and then intensified in the Ordovician.

Soils

If the rate of erosion is higher than the rate of soil formation the soils are being destroyed by erosion. Where soil is not destroyed by erosion, erosion can in some cases prevent the formation of soil features that form slowly. Inceptisols are common soils that form in areas of fast erosion.

While erosion of soils is a natural process, human activities have increased by 10-40 times the rate at which erosion is occurring globally. Excessive (or accelerated) erosion causes both "on-site" and "off-site" problems. On-site impacts include decreases in agricultural productivity and (on natural landscapes) ecological collapse, both because of loss of the nutrient-rich upper soil layers. In some cases, the eventual end result is desertification. Off-site effects include sedimentation of waterways and eutrophication of water bodies, as well as sediment-related damage to roads and houses. Water and wind erosion are the two primary causes of land degradation; combined, they are responsible for about 84% of the global extent of degraded land, making excessive erosion one of the most significant environmental problems.

Types of Erosion

Water Erosion

High intensity summer rainfalls represent a significant risk of erosion by water. Raindrops hit bare soil with enough force to break the soil aggregates. These fragments wash into soil pores and prevent water from infiltrating the soil. Water then accumulates on the surface and increases runoff which takes soil with it.

Well-structured soils are less prone to break up, and the impact of raindrops is minimised if the soil surface is protected by plant or litter cover. The vulnerability of soils to water erosion depends on:

- Rainfall intensity (erosivity): High intensity rainfall creates serious risk as heavy drops on bare soil causes the soil surface to seal.
- Nature of the soil (erodibility): Clay soils vary in their ability to withstand raindrop impact.
- Slope length: If a slope is long, water running down the slope becomes deeper and moves faster, taking more soil with it.
- Slope steepness: The speed of runoff increases on steep slopes, which increases the power of water to break off and carry soil particles.

Sheet and Rill erosion on a cultivated paddock.

Sheet and Rill Erosion

Hill slopes are prone to sheet erosion and rill erosion. The amount of hill slope erosion largely depends on how the land is used.

Sheet erosion occurs when a thin layer of topsoil is removed over a whole hillside paddock—and may not be readily noticed.

Rill erosion occurs when runoff water forms small channels as it concentrates down a slope. These rills can be up to 0.3m deep. If they become any deeper than 0.3m they are referred to as gully erosion.

Scalding

Scalded landscape, western Queensland.

Scalding can occur when wind and water erosion removes the top soil and exposes saline or sodic soils. Raindrop impact alone can result in large amounts of soil being moved. However water or wind moving over the surface will remove more soil, and contribute to sheet, rill and gully erosion.

Erosion also tends to remove the lighter, smaller soil particles first (such as clay and silt), leaving fine and coarse sand behind.

A combination of large amounts of fine sand and small amounts of clay at the surface means the soil tends to seal and set hard, which limits infiltration (water entering the soil).

Gully Erosion

Gully erosion happens when runoff concentrates and flows strongly enough to detach and move soil particles.

For example, a waterfall may form, with runoff picking up energy as it plunges over the gully head. Splashback at the base of the gully head erodes the subsoil and the gully eats its way up the slope.

Gullies may develop in watercourses or other places where runoff concentrates. In cultivation or pastures, advanced rill erosion can develop into gully erosion.

This type of erosion is highly visible and affects soil productivity, restricts land use, and can damage roads, fences and buildings.

Gully depth is often limited by the depth of the underlying rock which means gullies are normally less than 2m deep. However, gullies may reach depths of 10–15m on deep alluvial and colluvial soils.

A highly eroded gully.

Causes

Gully development may be triggered by:

- Cultivation or grazing on soils susceptible to gully erosion.
- Increased runoff from land use changes such as tree clearing in a catchment or construction of new residential areas.
- Runoff concentration caused by furrows, contour banks, waterways, dam by-washes, stock pads, fences, tracks or roads.
- Improper design, construction or maintenance of waterways in cropping areas.
- Poor vegetation cover that may be caused by overgrazing, fires or salinity problems.
- Low flows or seepage flows over a long period.
- 'Down cutting' in a creek that causes gullies to advance up the drainage lines flowing into it.
- Diversion of a drainage line to an area of high erosion risk, such as a steep creek bank or soil that is highly prone to erosion.

Tunnel Erosion

Tunnel erosion is the removal of subsoil.

When water penetrates through a soil crack or a hole where a root has decayed, the soil disperses and is carried away with the flow to leave a small tunnel.

Tunnel erosion in Marburg, Queensland.

Initially, the surface soil remains relatively intact but, with every flow, the tunnel becomes larger and the soil may eventually collapse and form a gully.

The whole process speeds up significantly if an outlet is provided (such as an existing gully or cutting in a roadside) as this allows free flow of subsurface drainage water.

Soils vulnerable to tunnel erosion have dispersible subsoils with naturally high levels of sodium. Such soils are referred to as being sodic and are called Sodosols.

When clods of these soils are exposed to water, they readily break down into individual particles of sand, silt and clay which are easily removed as water moves through the subsoil.

Stream Bank Erosion

Severe stream bank erosion.

Recent floods have made stream bank erosion a widespread problem across Queensland.

The major cause of stream bank erosion is the destruction of vegetation on river banks (generally by clearing, overgrazing, cultivation, vehicle traffic up and down banks or fire) and the removal of sand and gravel from the stream bed.

Stream bank erosion can also be accelerated by factors such as:

- Stream bed lowering or infill.
- Inundation of bank soils followed by rapid drops in flow after flooding.

- Saturation of banks from off-stream sources.
- Redirection and acceleration of flow around infrastructure, obstructions, debris or vegetation within.
- Soil characteristics such as poor drainage or seams of readily erodible material within the bank profile.
- Wave action generated by wind or boat wash.
- Intense rainfall events (e.g. cyclones).

Erosion on Floodplains

Erosive flooding on a floodplain.

Some of best agricultural land is on floodplains because of the high fertility soils and availability of water for irrigation. These areas are subjected to high velocity floods that erode soils with insufficient surface cover.

This erosive flooding can remove the entire layer of cultivated topsoil exposing compacted subsoils. It is common for such areas to be stripped of 0.1 to 0.15m of topsoil.

The risk of erosion on floodplains depends on:

- Flood velocities: The bigger the flood, the higher the velocities.
- The orientation of crop rows and the amount of protective cover provided by crops or stubble when flooding occurs.

Problems can also occur at the end of a flood event when slow moving water flows over saturated soils.

Mass Movement

Mass movement occurs on cleared slopes in coastal areas. Gravity moves earth, rock and soil material downslope both slowly (millimetres per year) and suddenly (e.g. rock falls).

Example of mass movement—landslip on a steep slope near Coolangatta, Queensland.

Different forms of mass movement include:

- Soil creep.
- Earthflow.
- Slumping.
- Landslips.
- Landslides.
- Rock avalanches.

During periods of prolonged and heavy rainfall, water entering permeable soils can be stopped by a barrier such as bedrock or a clay-rich soil horizon.

The heavy weight of this saturated soil can slide downslope if it is sitting on a rock surface loosened by the build-up of water in the soil.

How to Avoid

It is better to prevent landslides than pay for expensive, time-consuming rehabilitation which may only be partially effective.

Trees should not be cleared from susceptible locations.

On cleared areas, plant locally adapted, deep-rooted, fast-growing trees.

As excessive water intake is the most common trigger of landslip, avoid obstructions such as dams or cross-slope drains.

Wind Erosion

Wind erosion is a significant problem in the arid grazing lands.

It is most likely to occur when strong winds blow over light-textured soils that have been heavily grazed during drought periods.

Dust storm, western Queensland.

It contributes to scalding, a process that forms smooth, bare areas on impermeable subsoils.

These areas, which vary from a few square metres to hundreds of hectares, are difficult to revegetate due to:

- Lack of topsoil.
- Low permeability.
- Their often saline surface.

Generally, wind erosion is not a serious issue in cropping areas. Most soils cultivated have a heavy texture—forming relatively large aggregates that are too coarse to be carried by strong winds.

However, sandy soils are vulnerable to wind erosion because they cannot store very much moisture and have low fertility. This limits their use, with only small areas (where irrigation is available) being cultivated for horticulture or growing vegetables.

Consequences of Human-made Soil Erosion

Soil Retrogression and Degradation

Soil retrogression and degradation are two regressive evolution processes associated with the loss of equilibrium of a stable soil. Retrogression is primarily due to soil erosion and corresponds to a phenomenon where succession reverts the land to its natural physical state. Degradation is an evolution, different from natural evolution, related to the local climate and vegetation. It is due to the replacement of primary plant communities (known as climax vegetation) by the secondary communities. This replacement modifies the humus composition and amount, and affects the formation of the soil. It is directly related to human activity. Soil degradation may also be viewed as any change or ecological disturbance to the soil perceived to be deleterious or undesirable.

At the beginning of soil formation, the bare rock out crops is gradually colonized by pioneer species (lichens and mosses). They are succeeded by herbaceous vegetation, shrubs and finally forest. In parallel, the first humus-bearing horizon is formed (A horizon), followed by some mineral horizons (B horizons). Each successive stage is characterized by a certain association of soil/vegetation and environment, which defines an ecosystem.

Intensive tillage result on soil degradation.

Willow hedge strengthened with fascines for the limitation of runoff, northern France.

After a certain time of parallel evolution between the ground and the vegetation, a state of steady balance is reached. This stage of development is called climax by some ecologists and "natural potential" by others. Succession is the evolution towards climax. Regardless of its name, the equilibrium stage of primary succession is the highest natural form of development that the environmental factors are capable of producing.

The cycles of evolution of soils have very variable durations, between tens, hundreds, or thousands of years for quickly evolving soils (A horizon only) to more than a million years for slowly developing soils. The same soil may achieve several successive steady state conditions during its existence, as exhibited by the Pygmy forest sequence in Mendocino County, California. Soils naturally reach a state of high productivity, from which they naturally degrade as mineral nutrients are removed from the soil system. Thus older soils are more vulnerable to the effects of induced retrogression and degradation.

Ecological Factors Influencing Soil Formation

There are two types of ecological factors influencing the evolution of a soil (through alteration and humification). These two factors are extremely significant to explain the evolution of soils of short development.

- A first type of factor is the average climate of an area and the vegetation which is associated (biome).

- A second type of factor is more local, and is related to the original rock and local drainage. This type of factor explains appearance of specialized associations (ex peat bogs).

Biorhexistasy Theory

The destruction of the vegetation implies the destruction of evoluted soils, or a regressive evolution. Cycles of succession-regression of soils follow one another within short intervals of time (human actions) or long intervals of time (climate variations).

The climate role in the deterioration of the rocks and the formation of soils lead to the formulation of the theory of the biorhexistasy.

- In wet climate, the conditions are favorable to the deterioration of the rocks (mostly chemically), the development of the vegetation and the formation of soils; this period favorable to life is called biostasy.
- In dry climate, the rocks exposed are mostly subjected to mechanical disintegration which produces coarse detrital materials: this is referred to as rhexistasy.

Perturbations of the Balance of a Soil

When the state of balance, characterized by the ecosystem climax is reached, it tends to be maintained stable in the course of time. The vegetation installed on the ground provides the humus and ensures the ascending circulation of the matters. It protects the ground from erosion by playing the role of barrier (for example, protection from water and wind). Plants can also reduce erosion by binding the particles of the ground to their roots.

A disturbance of climax will cause retrogression, but often, secondary succession will start to guide the evolution of the system after that disturbance. Secondary succession is much faster than primary because the soil is already formed, although deteriorated and needing restoration as well.

However, when a significant destruction of the vegetation takes place (of natural origin such as an avalanche or human origin), the disturbance undergone by the ecosystem is too important. In this latter case, erosion is responsible for the destruction of the upper horizons of the ground, and is at the origin of a phenomenon of reversion to pioneer conditions. The phenomenon is called retrogression and can be partial or total (in this case, nothing remains beside bare rock). For example, the clearing of an inclined ground, subjected to violent rains, can lead to the complete destruction of the soil. Man can deeply modify the evolution of the soils by direct and brutal action, such as clearing, abusive cuts, forest pasture, litters raking. The climax vegetation is gradually replaced and the soil modified (example: replacement of leafy tree forests by moors or pines plantations). Retrogression is often related to very old human practices.

Influence of Human Activity

Soil erosion is the main factor for soil degradation and is due to several mechanisms: water erosion, wind erosion, chemical degradation and physical degradation.

Erosion is strongly related to human activity. For example, roads which increase impermeable surfaces lead to streaming and ground loss. Agriculture also accelerates soil erosion (increase of field size, correlated to hedges and ditches removal). Meadows are in regression to the profit of plowed lands. Spring cultures (sunflower, corn, beet) surfaces are increasing and leave the ground naked in winter. Sloping grounds are gradually colonized by vine. Lastly, use of herbicides leaves the ground naked between each crop. New cultural practices, such as mechanization also increases the risks of erosion. Fertilization by mineral manures rather than organic manure gradually destructure the soil. Many scientistsobserved a gradual decrease of soil organic matter content in soils, as well as a decrease of soil biological activity (in particular, in relation to chemical uses). Lastly, deforestation, in particular, is responsible for degradation of forest soils.

Agriculture increases the risk of erosion through its disturbance of vegetation by way of:

- Overgrazing of animals.
- Monoculture planting.
- Row cropping.
- Tilling or plowing.
- Crop removal.
- Land-use conversion.

Consequences of Soil Regression and Degradation

- Yields impact: Recent increases in the human population have placed a great strain on the world's soil systems. More than 6 billion people are now using about 38% of the land area of the Earth to raise crops and livestock. Many soils suffer from various types of degradation, that can ultimately reduce their ability to produce food resources. Slight degradation refers to land where yield potential has been reduced by 10%, moderate degradation refers to a yield decrease from 10-50%. Severely degraded soils have lost more than 50% of their potential. Most severely degraded soils are located in developing countries.
- Natural disasters: Natural disasters such as mud flows, floods are responsible for the death of many living beings each year.
- Deterioration of the water quality: The increase in the turbidity of water and the contribution of nitrogen and of phosphorus can result in eutrophication. Soils particles in surface waters are also accompanied by agricultural inputs and by

some pollutants of industrial, urban and road origin (such as heavy metals). The ecological impact of agricultural inputs (such as weed killer) is known but difficult to evaluate because of the multiplicity of the products and their broad spectrum of action.

- Biological diversity: Soil degradation may involve perturbation of microbial communities, disappearance of the climax vegetation and decrease in animal habitat, thus leading to a biodiversity loss and animal extinction.
- Economic loss: The estimated costs for land degradation are US $40 billion per year. This figure does not take into account negative externalities such as the increased use of fertilizers, loss of biodiversity and loss of unique landscapes.

Soil Enhancement, Rebuilding and Regeneration

Problems of soil erosion can be fought, and certain practices can lead to soil enhancement and rebuilding. Even though simple, methods for reducing erosion are often not chosen because these practices outweigh the short-term benefits. Rebuilding is especially possible through the improvement of soil structure, addition of organic matter and limitation of runoff. However, these techniques will never totally succeed to restore a soil (and the fauna and flora associated to it) that took more than 1000 years to build up. Soil regeneration is the reformation of degraded soil through biological, chemical, and or physical processes.

When productivity declined in the low-clay soils of northern Thailand, farmers initially responded by adding organic matter from termite mounds, but this was unsustainable in the long-term. Scientists experimented with adding bentonite, one of the smectite family of clays, to the soil. In field trials, conducted by scientists from the International Water Management Institute in cooperation with Khon Kaen University and local farmers, this had the effect of helping retain water and nutrients. Supplementing the farmer's usual practice with a single application of 200 kg bentonite per rai (6.26 rai = 1 hectare) resulted in an average yield increase of 73%. More work showed that applying bentonite to degraded sandy soils reduced the risk of crop failure during drought years.

In 2008, three years after the initial trials, IWMI scientists conducted a survey among 250 farmers in northeast Thailand, half who had applying bentonite to their fields and half who had not. The average output for those using the clay addition was 18% higher than for non-clay users. Using the clay had enabled some farmers to switch to growing vegetables, which need more fertile soil. This helped to increase their income. The researchers estimated that 200 farmers in northeast Thailand and 400 in Cambodia had adopted the use of clays, and that a further 20,000 farmers were introduced to the new technique.

Land Degradation

Land degradation is a process in which the value of the biophysical environment is affected by a combination of human-induced processes acting upon the land. It is viewed as any change or disturbance to the land perceived to be deleterious or undesirable. Natural hazards are excluded as a cause; however human activities can indirectly affect phenomena such as floods and bush fires.

Serious land degradation in Nauru after the depletion of the phosphate covers through mining.

This is considered to be an important topic of the 21st century due to the implications land degradation has upon agricultural productivity, the environment, and its effects on food security. It is estimated that up to 40% of the world's agricultural land is seriously degraded.

"About a quarter of the Earth's ice-free land area is subject to human-induced degradation (medium confidence). Soil erosion from agricultural fields is estimated to be currently 10 to 20 times (no tillage) to more than 100 times (conventional tillage) higher than the soil formation rate (medium confidence)."

Measures

There are four main ways of looking at land degradation and its impact on the environment around it:

- A temporary or permanent decline in the productive capacity of the land. This can be seen through a loss of biomass, a loss of actual productivity or in potential productivity, or a loss or change in vegetative cover and soil nutrients.
- Action in the land's capacity to provide resources for human livelihoods. This can be measured from a base line of past land use.
- Loss of biodiversity: A loss of range of species or ecosystem complexity as a decline in the environmental quality.
- Shifting ecological risk: Increased vulnerability of the environment or people to destruction or crisis. This is measured through a base line in the form of pre-existing risk of crisis or destruction.

A problem with defining land degradation is that what one group of people might view as degradation, others might view as a benefit or opportunity. For example, planting crops at a location with heavy rainfall and steep slopes would create scientific and environmental concern regarding the risk of soil erosion by water, yet farmers could view the location as a favourable one for high crop yields.

Different Types

Potato field with soil erosion.

In addition to the usual types of land degradation that have been known for centuries (water, wind and mechanical erosion, physical, chemical and biological degradation), four other types have emerged in the last 50 years:

- Pollution, often chemical, due to agricultural, industrial, mining or commercial activities.
- Loss of arable land due to urban construction, road building, land conversion, agricultural expansion, etc.
- Artificial radioactivity, sometimes accidental.
- Land-use constraints associated with armed conflicts.

Overall, more than 36 types of land degradation can be assessed. All are induced or aggravated by human activities, e.g. Soil erosion, Soil contamination, Soil acidification, sheet erosion, silting, aridification, salinization, urbanization, etc.

Causes

Overgrazing by livestock can lead to land degradation.

Land degradation is a global problem largely related to agricultural use, deforestation and Climate Change. Causes include:

- Land clearance, such as clearcutting and deforestation.
- Agricultural depletion of soil nutrients through poor farming practices.
- Livestock including overgrazing and overdrafting.
- Inappropriate irrigation and overdrafting.
- Urban sprawl and commercial development.
- Vehicle off-roading.
- Quarrying of stone, sand, ore and minerals.
- Increase in field size due to economies of scale, reducing shelter for wildlife, as hedgerows and copses disappear.
- Exposure of naked soil after harvesting by heavy equipment.
- Monoculture, destabilizing the local ecosystem.
- Dumping of non-biodegradable trash, such as plastics.
- Invasive Species.
- Climate Change.
- Loss of soil carbon.

Soil erosion in a wheat field near Pullman, US.

Overcutting of vegetation occurs when people cut forests, woodlands and shrublands—to obtain timber, fuelwood and other products—at a pace exceeding the rate of natural regrowth. This is frequent in semi-arid environments, where fuelwood shortages are often severe.

Overgrazing is the grazing of natural pastures at stocking intensities above the livestock carrying capacity; the resulting decrease in the vegetation cover is a leading cause of wind and water erosion. It is a significant factor in Afghanistan. The growing population pressure, during 1980–1990, has led to decreases in the already small areas of agricultural land per person in six out of eight countries (14% for India and 21% for Pakistan).

Population pressure also operates through other mechanisms. Improper agricultural practices, for instance, occur only under constraints such as the saturation of good lands under population pressure which leads settlers to cultivate too shallow or too steep soils, plough fallow land before it has recovered its fertility, or attempt to obtain multiple crops by irrigating unsuitable soils.

High population density is not always related to land degradation. Rather, it is the practices of the human population that can cause a landscape to become degraded. Populations can be a benefit to the land and make it more productive than it is in its natural state. Land degradation is an important factor of internal displacement in many African and Asian countries.

Severe land degradation affects a significant portion of the Earth's arable lands, decreasing the wealth and economic development of nations. As the land resource base becomes less productive, food security is compromised and competition for dwindling resources increases, the seeds of famine and potential conflict are sown.

Climate Change and Land Degradation

"Climate change exacerbates land degradation, particularly in lowlying coastal areas, river deltas, drylands and in permafrost areas (high confidence). Over the period 1961–2013, the annual area of drylands in drought has increased, on average by slightly more than 1% per year, with large inter-annual variability. In 2015, about 500 million people lived within areas which experienced desertification between the 1980s and 2000s. The highest numbers of people affected are in South and East Asia, the circum Sahara region including North Africa, and the Middle East including the Arabian Peninsula (low confidence). Other dryland regions have also experienced desertification. People living in already degraded or desertified areas are increasingly negatively affected by climate change (high confidence)."

Significant land degradation from seawater inundation, particularly in river deltas and on low-lying islands, is a potential hazard that was identified in a 2007 IPCC report.

As a result of sea-level rise from climate change, salinity levels can reach levels where agriculture becomes impossible in very low-lying areas.

Preventing and Managing Erosion

The three main principles to control erosion are to:

- Use land according to its capability.
- Protect the soil surface with some form of cover.
- Control runoff before it develops into an erosive force.

Land Capability

Soil erosion can be avoided by using land within its capability.

The land's position, soil type and slope determine how vulnerable it will be to erosion. It may not be suitable for agriculture, or suitable only for an activity which limits erosion.

There are a number of resources to help determine how land should be used to avoid erosion:

- For cropping lands: Land management field manuals map and describe the land types in many districts and provide advice on land use and management for each soil type.
- For grazing lands: Maps of soils and land types are available for most areas. These give graziers an indication of what soils their property may have and are a useful planning tool. Search for soils maps and reports in your region.

Using land according to its capability: the steeper slopes and shallower soils suitable for growing pastures, and the lower slopes and deeper soils suitable for growing crops.

Surface Cover and Runoff

Surface cover is a major factor to control erosion because it reduces the impact of raindrops falling on bare soils and wind removing soil particles. It also reduces the speed of water flowing over the land.

Erosion risk is significantly reduced when there is more than 30% soil cover. Total cover is achievable for many grazing and cropping systems.

Runoff concentrates as it flows downslope. By the time rivers draining large catchments reach the coast, they are usually just a few hundred meters wide. Even though surface cover encourages runoff to spread, runoff concentration is inevitable.

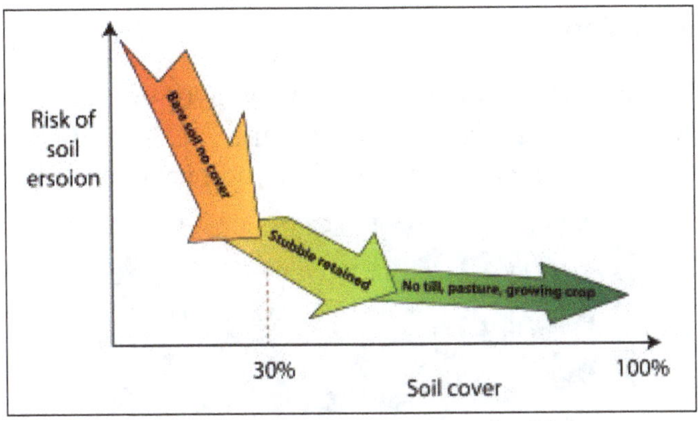

The relationship between cover and soil loss.

Coordination across the catchment is important when implementing runoff control measures. Runoff may pass through several properties and cross several roads (sometimes railway lines) as it passes from the most remote part of a catchment to a major drainage line or creek.

Experiment: The Importance of Cover

The importance of cover was shown in an experiment at Mt Mort, near Ipswich. In treatment C, where the land was almost bare, 70% of rainfall from a 54mm storm was lost as runoff. The soil loss from this one event was 22 tonnes per hectare. Treatments A and B with higher cover levels had much less runoff, soil and nutrient loss.

It can take thousands of years to form an inch of soil. The depth of soil lost from treatment C from this one storm event may take hundreds of years to replace, provided no further erosion occurs.

Table: Results from a 54mm storm at Mt Mort.

Treatment	A	B	C
Per cent cover	87	69	6
Total runoff from storm (mm)	1.5	14	38
Per cent of rainfall that ran off	3	26	70
Soil loss (t/ha)	0.03	0.3	22
Depth of soil lost (mm)	0.002	0.02	1.7
Sediment concentration (g/L)	1.5	1.9	63
Nitrogen removed (kg/ha)	0.14	1.9	15.3
Phosphorous removed (kg/ha)	0.02	0.26	4.3

Using Trees to Control Erosion

Trees are often considered to be the universal answer to control soil erosion. Tree roots help prevent landslides on steep slopes and stream bank erosion but they don't stop erosion on moderately sloping hill slopes.

In forests, the soil surface is usually protected by a layer of mulch from decaying vegetation as well as a variety of surface growing plants. If the soil is bare under the tree canopy from over grazing, vehicles or pedestrians, soil erosion will still occur.

Bare soil under trees is susceptible to erosion.

Erosion Control in Cropping Lands

Tillage

Conservation cropping practices that maintain cover on soils include minimum and zero tillage practices. Nowadays during the fallow period, farmers use tillage implements that kill weeds without burying stubble and herbicides to minimise the frequency of tillage.

Contour Banks and Strip Cropping

Runoff concentration is managed by structural measures such as contour banks in upland areas, or strip cropping on floodplains. These systems involve a total change in the way a farm is managed.

Runoff systems must be carefully planned. Flow between properties and across roads and railway lines must be coordinated and suit those affected by the changes.

Conservation cropping practices—a soybean crop planted directly into wheat stubble without any prior tillage.

When runoff water can impact neighbouring properties or infrastructure, land owners are encouraged to discuss with their neighbours and seek professional advice.

Approximately 80% of soil lost as a result of poor cover can be trapped in the paddock by contour banks. The banks channel the runoff at low speed into grassed waterways. Good surface cover between contour banks and in waterways will ensure their stability and dramatically reduce the amount of soil deposited in waterways.

Strip cropping over a floodplain on the Darling Downs.

On flood plains, strip cropping is used to spread flood flows rather than allowing it to concentrate.

Green Cane Harvesting

Another measure that maintains soil cover is green cane harvesting or 'trash blanketing'. When a cane crop is harvested, the leaves and tops of the cane are left on the ground as a 'trash blanket'. This protects the soil from erosion by raindrop impact. This practice has been widely adopted in many Queensland cane growing districts.

Well-maintained contour banks (left) and grassed waterways (right).

Green cane harvesting on a paddock in North Queensland.

Erosion Control in Grazing Lands

Surface Cover

Surface cover is the key to erosion control in grazing lands. It prevents erosion by maintaining the soil so it can absorb rainfall.

A well-managed pasture with good cover will ensure that runoff spreads rather than concentrates. Bore drains, tracks, roads, cattle pads and fences concentrate runoff, so careful planning is required to ensure that property improvements are located where they will not contribute to erosion.

The critical level of cover for pastures in tussock grasslands is about 40% cover and 1000 kilograms per hectare of dry grass. Ideally, this level of cover will exist at the beginning of the summer storm season.

The ideal stocking rate is flexible, and stock numbers should match available feed. Regular monitoring of pastures is necessary to achieve this. Long-term weather forecasting, using predictive tools such as the Southern Oscillation Index (SOI), has improved the options available for predicting droughts.

Opportunistic Spelling

Opportunistic spelling should also be part of a grazing strategy. A total spell in a good year may be required to allow desirable grasses to recover from past grazing. Grazing pressure can also be managed by locating watering points away from areas vulnerable to erosion.

Fire

Fire is useful for controlling woody weeds but it needs to be managed carefully. Regular burning of pastures will further reduce ground cover and promote runoff and erosion.

Use of a sediment fence and grassed strips on a development near Sinnamon Park, Brisbane.

Managing Erosion in Urban Areas

Queensland's rapidly increasing population and continued economic development require numerous construction projects and activities that expose soils to erosion.

The following approaches will help reduce erosion on development sites:

- Disturb minimal area when excavating.

- Where possible, divert upslope stormwater around the work site and other disturbed areas.
- Install sediment barriers (e.g. sediment fences or turf buffer strips) downslope of the building site to filter coarse sediments.
- Restrict vehicle access to one entry point where possible. Gravelling the access point will allow all weather access and minimise erosion.
- Connect a temporary or permanent downpipe to a stormwater system before laying the roof.
- Place all stockpiles on the construction site and behind a sediment barrier.
- Landscape all bare areas as soon as possible after construction is completed.

References

- Geomorphic-Weathering-309464790: researchgate.net, Retrieved 17 July, 2020
- Four-types-physical-weathering-6456598: sciencing.com, Retrieved 18 June, 2020
- Sims, G.K. 2014. "Soil degradation", in AccessScience@McGraw-Hill, http://www.accessscience.com, DOI 10.1036/1097-8542.757375
- 5-2-chemical-weathering: opentextbc.ca, Retrieved 15 March, 2020
- Toy, Terrence J.; Foster, George R.; Renard, Kenneth G. (2002). Soil erosion : processes, prediction, measurement, and control. New York: Wiley. ISBN 978-0-471-38369-7
- Examples-of-chemical-weathering-607608: thoughtco.com, Retrieved 14 February, 2020
- Hales, T. C.; Roering, Joshua (2007). "Climatic controls on frost cracking and implications for the evolution of bedrock landscapes". Journal of Geophysical Research: Earth Surface. 112 (F2): F02033. Bibcode:2007JGRF..112.2033H. CiteSeerX 10.1.1.716.110. doi:10.1029/2006JF000616

Landforms and Landscapes

5

- Landform
- Landscape
- Valley
- Fluvial Landscape
- Glacial and Glaciofluvial Landscapes
- Periglacial Landscapes
- Coastal Landscapes
- Tor
- Erosional Landforms
- Geomorphic System

A landform is a natural, geographical or artificial feature that appears on the Earth's solid surface. Landscape refers to the characteristic features of an area. Landscapes are divided into two categories – natural and human landscapes. This chapter closely examines landforms and landscapes to provide an extensive understanding of the subject.

Landform

Landforms are the natural features and shapes existent on the face of the earth. Landforms possess many different physical characteristics and are spread out throughout the planet. Together, landforms constitute a specific terrain and their physical arrangement in the landscape forms what is termed as topography. The physical features of landforms include slope, elevation, rock exposure, stratification and rock type.

Oceans and continents illustrate the largest grouping of landforms. They are they further subcategorized into many different landforms based on their physical features and shapes. Examples of distinctive landforms include mountains, valleys, plateaus, glaciers, hills, loess, deserts, shorelines, and plains. Features such as volcanoes, lakes, rivers, mid-ocean ridges, and the great ocean basins are also part of landform features.

Different Major Landforms on Earth

Major types of landforms on earth include mountains, valleys, plateaus, glaciers, hills, loess, plains and desserts.

Mountains

Mountains are lands physical features protruding high beyond the hills and very high up the land surface with steep top commonly shaped up to a peak. They are created through the action of incredible forces in the earth such as volcanic eruptions. Often, mountains occur in the ocean compared to land and some are seen as mountain islands as their peaks protrude out of the water. Mountain formation result from the forces of erosion, volcanism, or uplifts in the earth's crust.

The forces of heat and pressure within the earth's interior are the main influencing factors to these forces as stated by geologists. These forces can be summed up as the plate tectonic movements – theoretically defined as the division of the earth's outmost layer into several plates which are in constant motion. Hence, the uplifts are cause by collision or pulling apart of the plates that also triggers other various geologic activities such as the ejection of magma onto the surface or volcanic eruptions.

The movements also contribute to horizontal compression that is the deformation of crustal strata which gives rise to folds. The Himalayas and the Europe's Jura and Alps mountains are examples of mountains formed as a result of horizontal compression. Some mountain ranges are also formed as a result of wind, water, and ice erosion. Other mountains are created from volcanism.

Examples of volcanic mountains include Mount Fuji in Japan, Mount Vesuvius in Italy, Mount Erebus in Antarctica, and Mount Saint Helens in the United States. Majority of volcanic mountains have summit craters that still expel debris and steam.

Valleys

A valley is a lowland area or surface depression of the earth between higher lands such as mountains or hills. In simple terms, it can be defined as a natural trough bounded by mountains or hills on the surface of the earth sloping down to the lake, ocean or stream, which is created because of water or ice erosion. On this basis, the rivers or streams flowing through the valley empty the land's precipitation into the oceans.

The lowest parts of the valleys are very fertile and make very good farmlands. Majority of the valleys on land are made up of running streams and rivers and nearly all their floors slope downstream. Valleys within the mountains normally have narrow floors. The sides of a valley are termed as valley slopes or valley walls and the section of floor along riverbanks are referred to as flood plains.

Valleys physical features include U-shaped and V-shaped caused through the forces of erosion by the flowing masses that persistently widens and deepens the valley. The flowing masses are either water or glacier that carries away huge amounts of debris. Very narrow and deep valleys are known as canyons.

Plateaus

Plateaus are fairly flat areas higher than the land surrounding it. The surrounding areas may have very steep slopes. Some plateaus such as the Tibet are situated between mountain ranges. Plateaus cover wide land areas and together with their enclosed basins they cover approximately 45% of the entire earth's land surface.

Some plateaus, for instance the Columbia Plateau of the United States and the Deccan of India are basaltic and were created because of lava flows spreading to thousands of square kilometers thereby building up the fairly flat land surfaces. Other plateaus form as a result of upward folding while some are due to the erosion of the nearby land that leaves them elevated. Because plateaus are elevated, they are subject to erosion.

Low plateaus make up good farming regions whereas high plateaus are considered great for grazing livestock. Most of the world's high plateaus are deserts. Other typical examples of plateaus include the Bolivian plateau in South America, the Colorado plateau of the United States, the Laurentian Plateau and the plateaus of Iran, Arabia, and Anatolia.

Glaciers

Glaciers are the perennial ice sheets on the planet. They are huge masses of ice that slowly move over the land surface, predominant in high mountains and the cold Polar Regions. The very low temperatures in the regions are the enabling factor for the build-up of snow and densification into ice at depths of 15 meters or even more. Most glaciers have density thickness in the ranges of 91 to 3000 meters.

The movements begin when the compaction is so dense that it moves under the pressure of its weight. It is estimated that more than 75% of the world's fresh water is currently locked away in these frozen reservoirs. The glaciers include the Greenland Ice Sheet and the Antarctic Ice Sheet. The Antarctic Ice sheets outlet glaciers comprise the steep and extensively long and narrow depression Beordmore Glacier, which is one of the longest outlets in the world. The gradual rice in continental temperatures has seen the glacial density grow smaller owing to melting.

Hills

Hills are raised areas on the surface of the earth with distinctive summits, but are not as high as mountains. Hills are created as a result of accumulation of rock debris or sand deposited by wind and glaciers. They can also be created by faulting when the faults go slightly upwards. Hills are generally present in low mountain valleys and plains.

The Black Hills are the most known. Deep erosions of areas previously raised by the earth's crust disturbances carry most of the soil away leaving behind a hill. Human activities may also create hill when soils are dug and piled giant masses. Volcanic eruptions as well create hills after the eruption when the molten materials or lava cools and hardens in a pile.

Loess

Loess is a fine-grained unstratified accumulation of clay and silt deposited by the wind. It appears brown or yellowish in color and is brought about by past glacial activity in an area. In precise, it is sedimentary deposits of clay and silt mineral particles which take place on land in some parts of the world. The thickness of loess deposits are just a few meters and one of their basic feature is known as the 'cat steps'.

It's held together by few clay particles and is mostly composed of quartz crystals which readily slide against each other. This property makes it highly susceptible to erosion which leads to the 'cat steps' feature. Loess formed after the ice age when the glaciers covering a relatively large portion of the earth melted and was carried away, exposing the vast plains of mud.

Upon drying of the mud, the forces of wind blew away the mud and exposed sediments and eventually deposited them as silt in stacks on top of each other to create bold steep banks. Regions made of loess are witnessed in eastern China and the north-western region of the United States.

Plains

Plains are broad flat areas on the earth's surface stretching over a wide area. Plains are lower than the land in their surrounding and can be found both inland and along the coast. Coastal plains rise from the seal level up to the point they meet raised landforms such as plateaus or mountains. The Atlantic Coastal plain is a prime example of a substantially populated and fertile coastal plain.

On the other hand, inland plains are generally found at high altitudes. Thick forests normally flourish on plains in humid climates. A fairly large portion of plains are covered by grasslands, for instance, the Great Plains in the United States. Human populations prefer settling on plains because of the soil and the terrain which is good for farming and building settlements such as cities, residential areas, and transportation networks. Flood plains are also in this category and they are formed as a result of continuous accumulation of sand, silt, and mud when rivers overflow its banks.

Deserts

Deserts are the hot and dry areas of the world. They are the arid and semi-arid lands with little or no vegetation. Deserts constitute approximately 20% of the earth's total land cover and are distinguished by little or no rainfall. The deserts are divided into four major categories including the Semi-Arid Deserts, the Hot and Dry Deserts, the Cold Deserts, and the Coastal Deserts.

These deserts are located in different areas of the world. Deserts experience very high temperatures, less cloud cover, low humidity, low atmospheric pressure, and very little rain, which makes them have very little vegetation cover. The soil cover is also rocky and shallow and with very little organic matter and as such, it only supports a few plants adapted to the conditions.

Plants such as cacti and short shrubs are the ones adapted to the desert conditions because they can conserve water and tolerate the high temperatures. Animals in the deserts include insects, small carnivores, snakes, lizards, and birds adapted to survive with very little water. These animals hide during the day till nightfall to avoid the heat. An example of a desert is the Sahara of North Africa.

Landscape

A landscape is the visible features of an area of land, its landforms, and how they integrate with natural or man-made features. A landscape includes the physical elements of geophysically defined landforms such as (ice-capped) mountains, hills, water bodies such as rivers, lakes, ponds and the sea, living elements of land cover including indigenous vegetation, human elements including different forms of land use, buildings, and structures, and transitory elements such as lighting and weather conditions. Combining both their physical origins and the cultural overlay of human presence, often created over millennia, landscapes reflect a living synthesis of people and place that is vital to local and national identity.

A prairie: Badlands National Park, South Dakota.

Tropical rainforest, Fatu Hiva Island, Marquesas Islands, French Polynesia.

The character of a landscape helps define the self-image of the people who inhabit it and a sense of place that differentiates one region from other regions. It is the dynamic backdrop to people's lives. Landscape can be as varied as farmland, a landscape park or wilderness.

The Earth has a vast range of landscapes, including the icy landscapes of polar regions, mountainous landscapes, vast arid desert landscapes, islands, and coast-al landscapes, densely forested or wooded landscapes including past boreal forests and tropical rainforests, and agricultural landscapes of temperate and tropical regions. The activity of modifying the visible features of an area of land is referred to as landscaping.

Autumn landscape in Rybiniszki, Latvia, watercolor by Stanisław Masłowski.

A typical Dutch landscape in South Holland.

There are several definitions of what constitutes a landscape, depending on context. In common usage however, a landscape refers either to all the visible features of an area of land (usually rural), often considered in terms of aesthetic appeal, or to a pictorial representation of an area of countryside, specifically within the genre of landscape painting. When people deliberately improve the aesthetic appearance of a piece of land—by changing contours and vegetation, etc.—it is said to have been landscaped, though the result may not constitute a landscape according to some definitions.

The word landscape (landscipe or landscaef) arrived in England—and therefore into the English language—after the fifth century, following the arrival of the Anglo-Saxons; these terms referred to a system of human-made spaces on the land. The term landscape emerged around the turn of the sixteenth century to denote a painting whose primary subject matter was natural scenery. Land may be taken in its sense of something to which people belong (as in England being the land of the English). The suffix scape is equivalent to the more common English suffix ship. The roots of ship are etymologically akin to Old English sceppan or scyppan, meaning to shape. The suffix schaft is related to the verb schaffen, so that ship and shape are also etymologically linked. The modern form of the word, with its connotations of scenery, appeared in the late sixteenth century when the term landschap was introduced by Dutch painters who used it to refer to paintings of inland natural or rural scenery. The word landscape, first recorded in 1598, was borrowed from a Dutch painters' term. The popular conception of the landscape that is reflected in dictionaries conveys both a particular and a general meaning, the particular referring to an area of the Earth's surface and the general being that which can be seen by an observer. An example of this second usage can be found as early as 1662 in the Book of Common Prayer:

"Could we but climb where Moses stood, and view the landscape over. (General Hymns, verse 536)".

Kukle, Czech Republic.

There are several words that are frequently associated with the word landscape:

- Scenery: The natural features of a landscape considered in terms of their appearance, especially when picturesque: spectacular views of mountain scenery.
- Setting: In works of narrative (especially fictional), it includes the historical mo-

ment in time and geographic location in which a story takes place, and helps initiate the main backdrop and mood for a story.

- Picturesque: The word literally means "in the manner of a picture; fit to be made into a picture", and used as early as 1703, and derived from an Italian term pittoresco, "in the manner of a painter". Gilpin's Essay on Prints defined picturesque as "a term expressive of that peculiar kind of beauty, which is agreeable in a picture".
- A view: "A sight or prospect of some landscape or extended scene; an extent or area covered by the eye from one point" (OED).
- Wilderness: An uncultivated, uninhabited, and inhospitable region.
- Cityscape (also townscape): The urban equivalent of a landscape. In the visual arts a cityscape (urban landscape) is an artistic representation, such as a painting, drawing, print or photograph, of the physical aspects of a city or urban area.
- Seascape: A photograph, painting, or other work of art which depicts the sea, in other words an example of marine art.

Physical Landscape

Geomorphology: The Physical Evolution of Landscape

Geomorphology is the scientific study of the origin and evolution of topographic and bathymetric features created by physical or chemical processes operating at or near Earth's surface. Geomorphologists seek to understand why landscapes look the way they do, to understand landform history and dynamics and to predict changes through a combination of field observations, physical experiments and numerical modeling. Geomorphology is practiced within physical geography, geology, geodesy, engineering geology, archaeology and geotechnical engineering. This broad base of interests contributes to many research styles and interests within the field.

The surface of Earth is modified by a combination of surface processes that sculpt landscapes, and geologic processes that cause tectonic uplift and subsidence, and shape the coastal geography. Surface processes comprise the action of water, wind, ice, fire, and living things on the surface of the Earth, along with chemical reactions that form soils and alter material properties, the stability and rate of change of topography under the force of gravity, and other factors, such as (in the very recent past) human alteration of the landscape. Many of these factors are strongly mediated by climate. Geologic processes include the uplift of mountain ranges, the growth of volcanoes, isostatic changes in land surface elevation (sometimes in response to surface processes), and the formation of deep sedimentary basins where the surface of Earth drops and is filled with material eroded from other parts of the landscape. The Earth surface and its topography therefore are an intersection of climatic, hydrologic, and biologic action with geologic processes.

List of Different Types of Landscape

Desert, Plain, Taiga, Tundra, Wetland, Mountain, Mountain range, Cliff, Coast, Littoral zone, Glacier, Polar regions of Earth, Shrubland, Forest, Rainforest, Woodland, Jungle, Moors.

Panorama of the Chaîne des Puys from Puy de Dôme in winter. Massif Central, France. An example of how past volcanic activity shaped a landscape.

Landscape Ecology

Landscape ecology is the science of studying and improving relationships between ecological processes in the environment and particular ecosystems. This is done within a variety of landscape scales, development spatial patterns, and organizational levels of research and policy.

Landscape is a central concept in landscape ecology. It is, however, defined in quite different ways. For example: Carl Troll conceives of landscape not as a mental construct but as an objectively given 'organic entity', a "harmonic individuum of space". Ernst Neef defines landscapes as sections within the uninterrupted earth-wide interconnection of geofactors which are defined as such on the basis of their uniformity in terms of a specific land use, and are thus defined in an anthropocentric and relativistic way.

A landscape is a heterogeneous land area composed of a cluster of interacting ecosystems that is repeated in similar form throughout, whereby they list woods, meadows, marshes and villages as examples of a landscape's ecosystems, and state that a landscape is an area at least a few kilometres wide. John A. Wiens opposes the traditional view expounded by Carl Troll, Isaak S. Zonneveld, Zev Naveh, Richard T. T. Forman/Michel Godron and others that landscapes are arenas in which humans interact with their environments on a kilometre-wide scale; instead, he defines 'landscape'—regardless of scale—as "the template on which spatial patterns influence ecological processes". Some define 'landscape' as an area containing two or more ecosystems in close proximity.

Landscape Science

The discipline of landscape science has been described as "bringing landscape ecology and urban ecology together with other disciplines and cross-disciplinary fields to identify patterns and understand social-ecological processes influencing landscape change". A paper entitled "Geography and landscape science" states that "The whole of the disciplines involved in landscape research will be referred to as landscape science, although this term was used first in 1885 by the geographers Oppel and Troll". A editorial defines landscape science as "research that seeks to understand the relationship between people and their environment, with a focus on land use change and data pertaining to land resources at the landscape scale". Landscape science as "the branch of physical geography that deals with natural territorial complexes (or geographic complexes, geosystems) as structural parts of the earth's geographic mantle" and states that "The basis of landscape science is the theory that the geographic landscape is the primary element in the physicogeo-graphical differentiation of the earth. Landscape science deals with the origin, structure, and dynamics of landscapes, the laws of the development and arrangement of landscapes, and the transformation of landscapes by the economic activity of man.", and asserts that it was founded in Russia in the early 20th century by L. S. Berg and others, and outside Russia by the German S. Passarge.

Integrated Landscape Management

Integrated landscape management is a way of managing a landscape that brings together multiple stakeholders, who collaborate to integrate policy and practice for their different land use objectives, with the purpose of achieving sustainable landscapes. It recognises that, for example, one river basin can supply water for towns and agriculture, timber and food crops for smallholders and industry, and habitat for biodiversity; the way in which each one of these sectors pursues its goals can have impacts on the others. The intention is to minimise conflict between these different land use objectives and ecosystem services. This approach draws on landscape ecology, as well as many related fields that also seek to integrate different land uses and users, such as watershed management.

Proponents of integrated landscape management argue that it is well-suited to address complex global challenges, such as those that are the focus of the Sustainable Development Goals. Integrated landscape management is increasingly taken up at the national, local and international level, for example the UN Environment Programme states that "UNEP champions the landscape approach de facto as it embodies the main elements of integrated ecosystem management".

Landscape Archaeology

Landscape archaeology or landscape history is the study of the way in which humanity has changed the physical appearance of the environment - both present and past. Landscape generally refers to both natural environments and environments constructed by human beings.

Medieval Ridge and Furrow above Wood Stanway, Gloucestershire, England.

Natural landscapes are considered to be environments that have not been altered by humans in any shape or form. Cultural landscapes, on the other hand, are environments that have been altered in some manner by people (including temporary structures and places, such as campsites, that are created by human beings). Among archaeologists, the term landscape can refer to the meanings and alterations people mark onto their surroundings. As such, landscape archaeology is often employed to study the human use of land over extensive periods of time. Landscape archaeology can be summed up by Nicole Branton's statement:

> "The landscapes in landscape archaeology may be as small as a single household or garden or as large as an empire", and "although resource exploitation, class, and power are frequent topics of landscape archaeology, landscape approaches are concerned with spatial, not necessarily ecological or economic, relationships. While similar to settlement archaeology and ecological archaeology, landscape approaches model places and spaces as dynamic participants in past behavior, not merely setting (affecting human action), or artifact (affected by human action)".

Cultural Landscape

The concept of cultural landscapes can be found in the European tradition of landscape painting. From the 16th century onwards, many European artists painted landscapes in favor of people, diminishing the people in their paintings to figures subsumed within broader, regionally specific landscapes.

The geographer Otto Schlüter is credited with having first formally used "cultural landscape" as an academic term in the early 20th century. In 1908, Schlüter argued that by defining geography as a Landschaftskunde (landscape science) this would give geography a logical subject matter shared by no other discipline. He defined two forms of landscape: the Urlandschaft (translation original landscape) or landscape that existed before major human induced changes and the Kulturlandschaft (translation 'cultural landscape') a landscape created by human culture. The major task of geography was to trace the changes in these two landscapes.

The Batad rice terraces, The Rice Terraces of the Philippine Cordilleras, the first site to be included in the UNESCO World Heritage List cultural landscape category in 1995.

It was Carl O. Sauer, a human geographer, who was probably the most influential in promoting and developing the idea of cultural landscapes. Sauer was determined to stress the agency of culture as a force in shaping the visible features of the Earth's surface in delimited areas. Within his definition, the physical environment retains a central significance, as the medium with and through which human cultures act. His classic definition of a 'cultural landscape' reads as follows:

> "The cultural landscape is fashioned from a natural landscape by a cultural group. Culture is the agent, the natural area is the medium, the cultural landscape is the result".

A cultural landscape, as defined by the World Heritage Committee, is the "cultural properties that represent the combined works of nature and of man."

The World Heritage Committee identifies three categories of cultural landscape, ranging from (i) those landscapes most deliberately 'shaped' by people, through (ii) full range of 'combined' works, to (iii) those least evidently 'shaped' by people (yet highly valued). The three categories extracted from the Committee's Operational Guidelines, are as follows:

- "A landscape designed and created intentionally by man".
- An "organically evolved landscape" which may be a "relict (or fossil) landscape" or a "continuing landscape".
- An "associative cultural landscape" which may be valued because of the "religious, artistic or cultural associations of the natural element".

Human Conceptions and Representations of Landscape

Landscape Gardens

The Chinese garden is a landscape garden style which has evolved over three thousand years. It includes both the vast gardens of the Chinese emperors and members of the Imperial Family, built for pleasure and to impress, and the more intimate gardens created by scholars, poets, former government officials, soldiers and merchants, made for reflection and escape from the outside world. They create an idealized miniature landscape, which is meant to express the harmony that should exist between man and nature. A typical Chinese garden is enclosed by walls and includes one or more ponds, scholar's rocks, trees and flowers, and an assortment of halls and pavilions within the garden, connected by winding paths and zig-zag galleries. By moving from structure to structure, visitors can view a series of carefully composed scenes, unrolling like a scroll of landscape paintings.

Stourhead garden, Wiltshire, England.

The English landscape garden, also called English landscape park or simply the 'English garden', is a style of parkland garden intended to look as though it might be a natural landscape, although it may be very extensively re-arranged. It emerged in England in the early 18th century, and spread across Europe, replacing the more formal, symmetrical jardin à la française of the 17th century as the principal style for large parks and gardens in Europe. The English garden (and later French landscape garden) presented an idealized view of nature. It drew inspiration from paintings of landscapes by Claude Lorraine and Nicolas Poussin, and from the classic Chinese gardens of the East, which had recently been described by European travellers and were realized in the Anglo-Chinese garden, and the philosophy of Jean-Jacques Rousseau.

Jichang Garden in Wuxi.

The English garden usually included a lake, sweeps of gently rolling lawns set against groves of trees, and recreations of classical temples, Gothic ruins, bridges, and other picturesque architecture, designed to recreate an idyllic pastoral landscape. The work of Lancelot "Capability" Brown and Humphry Repton was particularly influential. By the end of the 18th century the English garden was being imitated by the French landscape garden, and as far away as St. Petersburg, Russia, in Pavlovsk, the gardens of the future Emperor Paul. It also had a major influence on the form of the public parks and gardens which appeared around the world in the 19th century.

Landscape Architecture

Central Park, designed by Frederick Law Olmsted.

Landscape architecture is a multi-disciplinary field, incorporating aspects of botany, horticulture, the fine arts, architecture, industrial design, geology and the earth sciences, environmental psychology, geography, and ecology. The activities of a landscape architect can range from the creation of public parks and parkways to site planning for campuses and corporate office parks, from the design of residential estates to the design of civil infrastructure and the management of large wilderness areas or reclamation of degraded landscapes such as mines or landfills. Landscape architects work on all types of structures and external space – large or small, urban, suburban and rural, and with "hard" (built) and "soft" (planted) materials, while paying attention to ecological sustainability.

For the period before 1800, the history of landscape gardening (later called landscape architecture) is largely that of master planning and garden design for manor houses, palaces and royal properties, religious complexes, and centers of government. An example is the extensive work by André Le Nôtre at Vaux-le-Vicomte and at the Palace of Versailles for King Louis XIV of France. The first person to write of making a landscape was Joseph Addison in 1712. The term landscape architecture was invented by Gilbert Laing Meason in 1828 and was first used as a professional title by Frederick Law Olmsted in 1863. During the latter 19th century, the term landscape architect became used by professional people who designed landscapes. Frederick Law Olmsted used the term 'landscape architecture' as a profession for the first time when designing Central Park, New York City, US. Here the combination of traditional landscape gardening and the emerging field of city planning gave landscape architecture its unique focus. This use of the term landscape architect became established after Frederick Law Olmsted, Jr. and others founded the American Society of Landscape Architects (ASLA) in 1899.

Landscape Art

Landscape Photography

The Tetons and the Snake River.

Many landscape photographs show little or no human activity and are created in the pursuit of a pure, unsullied depiction of nature devoid of human influence, instead featuring subjects such as strongly defined landforms, weather, and ambient light. As with most forms of art, the definition of a landscape photograph is broad, and may include urban settings, industrial areas, and nature photography. Notable landscape photographers include Ansel Adams, Galen Rowell, Edward Weston, Ben Heine, Mark Gray and Fred Judge.

Landscape Painting

Salomon van Ruisdael.

The earliest forms of art around the world depict little that could really be called landscape, although ground-lines and sometimes indications of mountains, trees or other natural features are included. The earliest "pure landscapes" with no human figures are frescos from Minoan Greece of around 1500 BCE. Hunting scenes, especially those set in the enclosed vista of the reed beds of the Nile Delta from Ancient Egypt, can give a strong sense of place, but the emphasis is on individual plant forms and human and animal figures rather than the overall landscape setting. For a coherent depiction of a whole landscape, some rough system of perspective, or scaling for distance, is needed, and this seems from literary evidence to have first been developed in Ancient Greece in the Hellenistic period, although no large-scale examples survive. More ancient Roman landscapes survive, from the 1st century BCE onwards, especially frescos of landscapes decorating rooms that have been preserved at archaeological sites of Pompeii, Herculaneum and elsewhere, and mosaics.

The Chinese ink painting tradition of shan shui ("mountain-water"), or "pure" landscape, in which the only sign of human life is usually a sage, or a glimpse of his hut, uses sophisticated landscape backgrounds to figure subjects, and landscape art of this period retains a classic and much-imitated status within the Chinese tradition.

Both the Roman and Chinese traditions typically show grand panoramas of imaginary landscapes, generally backed with a range of spectacular mountains – in China often with waterfalls and in Rome often including sea, lakes or rivers. These were frequently used to bridge the gap between a foreground scene with figures and a distant panoramic vista, a persistent problem for landscape artists.

A major contrast between landscape painting in the West and East Asia has been that while in the West until the 19th century it occupied a low position in the accepted hierarchy of genres, in East Asia the classic Chinese mountain-water ink painting was traditionally the most prestigious form of visual art. However, in the West, history painting came to require an extensive landscape background where appropriate, so the theory did not entirely work against the development of landscape painting – for several centuries landscapes were regularly promoted to the status of history painting by the addition of small figures to make a narrative scene, typically religious or mythological.

Dutch Golden Age painting of the 17th century saw the dramatic growth of landscape painting, in which many artists specialized, and the development of extremely subtle realist techniques for depicting light and weather. The popularity of landscapes in the Netherlands was in part a reflection of the virtual disappearance of religious painting in a Calvinist society, and the decline of religious painting in the 18th and 19th centuries all over Europe combined with Romanticism to give landscapes a much greater and more prestigious place in 19th-century art than they had assumed before.

In England, landscapes had initially been mostly backgrounds to portraits, typically suggesting the parks or estates of a landowner, though mostly painted in London by an artist who had never visited the site. The English tradition was founded by Anthony van Dyck and other, mostly Flemish, artists working in England. By the beginning of the 19th century the English artists with the highest modern reputations were mostly dedicated landscapists, showing the wide range of Romantic interpretations of the English landscape found in the works of John Constable, J.M.W. Turner and Samuel Palmer. However all these had difficulty establishing themselves in the contemporary art market, which still preferred history paintings and portraits.

Thomas Cole "The Course of Empire The Arcadian or Pastoral State".

In Europe, as John Ruskin said, and Sir Kenneth Clark confirmed, landscape painting was the "chief artistic creation of the nineteenth century", and "the dominant art", with the result that in the following period people were "apt to assume that the appreciation of natural beauty and the painting of landscape is a normal and enduring part of our spiritual activity".

The Romantic Movement intensified the existing interest in landscape art, and remote and wild landscapes, which had been one recurring element in earlier landscape art, now became more prominent. The German Caspar David Friedrich had a distinctive style, influenced by his Danish training. To this he added a quasi-mystical Romanticism. French painters were slower to develop landscape painting, but from about the 1830s Jean-Baptiste-Camille Corot and other painters in the Barbizon School established a French landscape tradition that would become the most influential in Europe for a century, with the Impressionists and Post-Impressionists for the first time making landscape painting the main source of general stylistic innovation across all types of painting.

In the United States, the Hudson River School, prominent in the middle to late 19th century, is probably the best-known native development in landscape art. These painters created works of mammoth scale that attempted to capture the epic scope of the landscapes that inspired them. The work of Thomas Cole, the school's generally acknowledged founder, has much in common with the philosophical ideals of European landscape paintings — a kind of secular faith in the spiritual benefits to be gained from the contemplation of natural beauty. Some of the later Hudson River School artists, such as Albert Bierstadt, created less comforting works that placed a greater emphasis (with a great deal of Romantic exaggeration) on the raw, even terrifying power of nature. The best examples of Canadian landscape art can be found in the works of the Group of Seven, prominent in the 1920s. Emily Carr was also closely associated with the Group of Seven, though was never an official member. Although certainly less dominant in the period after World War I, many significant artists still painted landscapes in the wide variety of styles exemplified by Neil Welliver, Alex Katz, Milton Avery, Peter Doig, Andrew Wyeth, David Hockney and Sidney Nolan.

The term neo-romanticism is applied in British art history, to a loosely affiliated school of landscape painting that emerged around 1930 and continued until the early 1950s.

These painters looked back to 19th-century artists such as William Blake and Samuel Palmer, but were also influenced by French cubist and post-cubist artists such as Pablo Picasso, André Masson, and Pavel Tchelitchew. This movement was motivated in part as a response to the threat of invasion during World War II. Artists particularly associated with the initiation of this movement included Paul Nash, John Piper, Henry Moore, Ivon Hitchens, and especially Graham Sutherland. A younger generation included John Minton, Michael Ayrton, John Craxton, Keith Vaughan, Robert Colquhoun, and Robert MacBryde.

Valley

Valley is the elongate depression of the Earth's surface valleys are most commonly drained by rivers and may occur in a relatively flat plain or between ranges of hills or mountains. Those valleys produced by tectonic action are called rift valleys. Very narrow, deep valleys of similar appearance are called gorges. Both of these latter types are commonly cut in flat-lying strata but may occur in other geological situations.

The rugged Atlas Mountains surround a valley in Morocco.

Wherever sufficient rainfall occurs, opportunity exists for the land surface to evolve to the familiar patterns of hills and valleys. There are, of course, hyperarid environments where fluvial activity is minimal. There also are geomorphological settings where the permeability of rocks or sediments induce so much infiltration that water is unable to concentrate on the land surface. Moreover, some landscapes may be so young that insufficient time has elapsed for modification by fluvial action.

Probably the world's deepest subaerial valley is that of the Kāli Gandaki River in Nepal. Lying between two 8,000-metre (26,000-foot) Himalayan peaks, Dhaulāgiri and Annapūrna, the valley has a total relief of six kilometres (four miles). Because the Himalayas are one of the Earth's most active areas of tectonic uplift, this valley well illustrates the principle that the most rapid downcutting occurs in areas of the most rapid uplift. The reason for this seeming paradox lies in the energetics of the processes of degradation that characterize valley formation. The steeper the gradient or slope of a stream, the greater its expenditure of power on the streambed. Thus, as uplift creates higher relief and steeper slopes, rivers achieve greater power for erosion. As a consequence, the most rapid processes of relief reduction can occur in areas of most rapid relief production.

Perhaps the most famous example of a canyon is the Grand Canyon of the Colorado River in northern Arizona. The Grand Canyon is about 1.6 km (1 mile) deep and 180 metres (590 feet) to 30 kms (19 miles) wide and occurs along a 443 kms (275 miles) long reach where the Colorado River incised into a broad upwarp of sedimentary rocks.

Geomorphic Characteristics

The relief of valleys and canyons is produced by the incising action of rivers. Hillslope processes are indeed critical in the development of valley sides, but it is rivers that lower the level of erosion through degradation. Rivers ultimately adjust to a baselevel, defined as the lowest point at which potential energy can be transformed to the kinetic energy of river flow. In most cases, the ultimate baselevel for rivers is sea level. Some rivers drain to enclosed basins below sea level, as, for example, the Jordan River, which flows to the Dead Sea in Israel and Jordan. Moreover, rivers may adjust to local baselevels, including zones of resistance to incision, lakes, and dams (both natural and artificial).

Valley Longitudinal Profiles

The longitudinal profile of a valley is the gradient throughout its length. Valleys formed by river action typically have a concave upward profile, steep in the headwaters and gentle in the lower reaches. The lower end of such a profile is adjusted to an effective lower limit of erosion defined by the baselevel.

In an ideal case of river adjustment to uniformly resistant materials, the longitudinal profile of a stream assumes a characteristic form that minimizes variations in transporting power. Power in a river derives from the rate of transfer of potential energy, dE/dt, which depends on the rate of fall in elevation of water, dy/dt, according to:

$$\frac{dE}{dt} = mg\frac{dy}{dt}$$

where E is energy, t is time, m is mass, g is the acceleration of gravity, and y is elevation. The rate of fall in elevation, in turn, can be expressed as follows:

$$\frac{dy}{dt} = \frac{dy}{dx}\frac{dx}{dt} = SV,$$

where S is the slope (fall in elevation, dy, with downstream horizontal distance, dx) and V is the flow velocity (change in horizontal distance, dx, with time, dt).

Combining above equations and using the fluid density ρ (mass per unit volume of water), one obtains:

$$\frac{dE}{dt} = \rho g(W.D.L)SV,$$

where W is channel width, D is channel depth, L is a unit length of stream, and the other parameters are as defined above. Because flow discharge Q is defined as:

$$Q = W.D.V,$$

The power per unit length of flow, Ω, can be expressed as:

$$\Omega = dE / dt / L = \rho gQS.$$

It should be noted that in order to minimize variation in power, a river increasing its discharge in a downstream direction must decrease its slope. Thus, slope must be constantly decreasing downstream, explaining the concave upward character of the longitudinal profile.

The idealized concave upward longitudinal profile defined purely by energy considerations, noted above, only occurs where channel bed resistances and adequate adjustment time permit. Resistant zones of bedrock require greater power for a stream to incise at a given discharge Q than do less resistant zones. Therefore, by equation ($\Omega = dE / dt / L = \rho gQS.$) the stream gradient S must be locally steeper at resistant zones. Similarly, a rapid base-level change, such as a fall of sea level, may not allow adequate time for the entire longitudinal profile to adjust. One indication of such effects on a longitudinal profile is a nick point, or abrupt change in slope of the profile.

Valley Cross Profiles

The cross profiles of valleys involve a combination of fluvial and hillslope processes. Although slopes and rivers are often studied separately by process geomorphologists, hills and valleys are the features that dominate landscapes. In upland areas cross profiles of valleys are often narrow and deep. Canyon morphologies are most common. Further downstream, valley floors are wider and often dominated by floodplains and terraces.

Types of Valleys

One of the few classifications of valleys is that used by the German climatic geomorphologists Herbert Louis and Julius Büdel. In areas of rapid uplift and intense fluvial action such as tropical mountains, Kerbtal (German for "notched valley") forms occur. These are characterized by steep, knife-edge ridges and valley slopes meeting in a V-shape. Where slopes are steep but a broad valley floor occurs, Sohlenkerbtal (meaning precisely a valley with such characteristics) is the prevailing form. Valleys of this kind develop under the influence of groundwater flow in Hawaii. Gutter-shaped valleys with convex sides and broad floors are called Kehltal; and broad, flat valleys of planation surfaces are termed Fachmuldental.

It is important to remember that the form of valleys reflects not only modern processes but also ancient ones. The entire valley or some landforms within it may be relict, with features inherited from past geologic periods during which occurred tectonic and climatic processes of intensities quite different from those prevailing today.

Hillslopes

Hillslopes constitute the flanks of valleys and the margins of eroding uplands. They are the major zones where rock and soil are loosened by weathering processes and then transported down gradient, often to a river channel.

Two major varieties of hillslopes occur in nature. On weathering-limited slopes, transport processes are so efficient that debris is removed more quickly than it can be generated by further weathering. Such hillslopes develop a faceted or angular morphology in which an upper free face, or cliff, contributes debris to a lower slope of accumulation. Slopes of this sort are especially common on bare rock where the profile of the slope is determined by the resistance of the rock, not by the erosional processes acting on it. One consequence of this is that many rock slopes retreat parallel to themselves in order to preserve the characteristic slope angle for a rock type of given strength. If the features of the rock change with depth into the slope, however, the characteristic angle of the slope will change. Rock slopes develop where weathering and soil erosion are slow (as in arid regions) and where rock resistance is high.

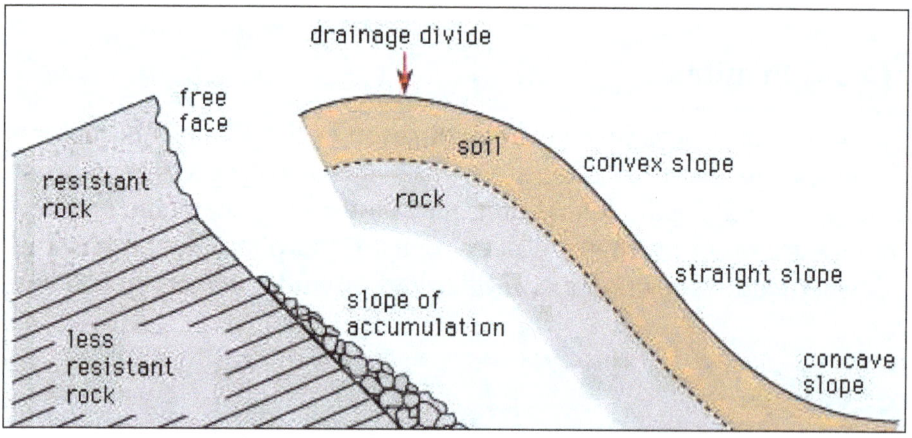

Comparison of idealized profiles for weathering-limited, faceted hillslopes (left) and transport-limited, sigmoid hillslopes (right).

The second major variety of slope is transport limited. Transport-limited slopes occur where weathering processes are efficient at producing debris but where transport processes are inefficient at removing it from the slope. Such slopes lack free faces and faceted appearances, and they are generally covered with a soil mantle. The profile of this type of slope generally has a sigmoid appearance, with convex, straight, and concave segments. The shape of the slope is an expression of the process acting upon it.

Convex slope segments commonly occur in the upper parts of soil-mantled slopes, as near the drainage divide. The noted American geomorphologist G.K. Gilbert elucidated the principles applying to convex slopes in his study of piles of mining-waste debris in California. The processes of soil creep and raindrop splash erode soil on the upper parts of slopes. Since soil eroded from the upper slope must pass each point below it, the volume of soil moved increases with distance from the divide. Since the transport rate for creep and rain splash is proportional to the slope angle, the slope angle must also increase from the divide, resulting in the slope convexity.

Straight slope segments are dominated by mass movement processes. Talus slopes are a type in which debris piles up to a characteristic angle of repose. When new debris is added to the slope, thereby locally increasing the angle, the slope adjusts by movement of the debris to re-establish the angle. Again, the result is a dynamic equilibrium in which the landform adjusts to processes acting upon it.

Concave slopes are especially common where overland-flow runoff transports sediment derived from upper slopes. Because the collection area for wash increases downslope and discharge Q is proportional to collection area, stream power—equation $\Omega = dE / dt / L = \rho gQS$ —can be maintained at lower slope angles. In addition, the size of particles being transported decreases downslope because of weathering and abrasion. Because the finer particles are easier to transport, slope angles can be reduced in the downslope direction. The result is a concave shape to the slope profile.

River valleys figure prominently in the evolutionary sequence of landscape development conceived by W.M. Davis. Unfortunately Davis' marvelous deductive scheme of progressive landscape change with time was somewhat abused by those who employed it merely for description and classification. By the mid-20th century, the focus of geomorphological research shifted from evolutionary sequences to studies of processes. Today, new procedures for radiometric dating have rekindled interest in long-term landscape evolution.

Valley development with time can be conceived of as a functional relationship, as follows:

$$v = f(c, r, l, p, t)$$

where v is the valley morphology, c is the climate, r is relief factors including slope, l is lithology and rock structure, p is the type of process operating (surface runoff or spring sapping), and t is time.

Valley morphology can be described in numerous ways. A useful measure is drainage density Dd, which relates the length of valleys (or streams) L to the area A in which they occur:

$$Dd = \frac{\Sigma L}{A}$$

In many applications, A is defined as the drainage area in which a network of valleys is developed. There is a close relationship of drainage density to hillslope angles and local relief. For a given relief, higher drainage density results in short, steep valley-side slopes. For the same relief, a lower drainage density results in long, gentle slopes.

Processes

Runoff Processes

When rain falls on a land surface, part of it may infiltrate, depending on the rate of rainfall and the permeability of the substrate. The amount of rainfall that exceeds the infiltration capacity collects in pools and eventually flows over the land surface. This process of overland flow is quite inefficient because a large surface area greatly resists water movement. Depending on the substrate resistance and power of the flow, the tendency is to incise to form a channel. This transition from overland flow to channel flow is the first step toward a response to rainfall input. Eventually the dissection by channels leads to the differentiation of hills from valleys.

Not all the rainfall is transformed to overland flow and infiltration to groundwater. A portion is lost to evaporation and to transpiration by plants. What eventually flows off the landscape from surface and subsurface sources is the runoff R given by:

$$R = P - ET \pm S,$$

where P is the precipitation and ET is the combination of evaporation and transpiration; S is a storage term for water held in plants, soils, and subsurface rocks. The overland flow component of runoff appears very quickly after storms, while the subsurface flow components appear much more slowly. In channels, all forms of runoff generate increased stream power because of increased discharge. This allows streams to incise, thereby deepening valleys, which may widen through hillslope processes.

Sapping

Sapping is a process of hillslope or scarp recession by the undermining of an overlying resistant material in the form of weathering or water flow occurring in an underlying less-resistant material. A variation of this process, spring sapping, occurs where groundwater outflow undermines slopes and, where appropriately concentrated, contributes to the development of valleys. The action of groundwater in sapping may be concentrated at valley heads, leading to headward growth. Both enhanced weathering and direct erosion by the concentrated fluid flow lead to slope undermining and collapse at sites of groundwater outflow.

A conceptual model of valley development by sapping can be envisioned with the initial condition of a water table having a regional slope toward a hydraulic sink provided by a depressed region. Water emerging along a spring line would then foster chemical weathering and thereby increase the porosity of the seepage zone, reducing the local rock tensile strength and rendering the weathering zone more susceptible to erosional undercutting of adjacent slopes. Local zones of heterogeneity in the rock will result in some zones achieving the critical conditions necessary for such undermining before other zones achieve them. Joints, faults, and folds serve this function. These critical zones then experience enhanced undermining. Once initiated, this process becomes self-enhancing because the lines of groundwater flow converge on the spring head. The increased flow accelerates chemical weathering, which leads to further piping at the same site.

The farther a spring head retreats, the greater the flow convergence that it generates, thereby increasing the rate of headward erosion. Headward sapping proceeds faster than valley widening because the valley head is the site of greatest flow convergence. Headward growth, however, may intersect other zones that are highly susceptible to sapping. A particularly favourable zone will result in a tributary that also experiences headward growth and that may generate tributaries of its own. Thus, sapping that occurs in a zone of jointing or faulting will develop a pattern aligned with those structures. It will, however, be organized by the hydraulic controls on the groundwater flow.

This process of sapping, headward retreat, and branching eventually forms a network of valleys. The developing network works to counteract the self-enhancing effect of flow concentration. As spring heads migrate to the neighbourhood of one another, their demands for the available groundwater compete with each other. Eventually an equilibrium is achieved at some optimum drainage density.

Excellent examples of valleys formed by sapping are found in the massive sandstone terrains of the Colorado Plateau. Groundwater seepage from the sandstone contributes to local disintegration of the bedrock at the bases of cliffs, thereby undermining slopes and leading to backwearing. Because of structural concentration of water flow along joints and faults, valleys grow headward along zones of structural weakness. Canyons formed by sapping have prominent structural control vertical to overhanging walls, flat floors, elongate shape, low drainage density (leaving undissected uplands), relatively short tributaries to main trunk valleys, irregular variation in valley width as a function of valley length, and theatre-like valley heads. Many of the sapping valleys of the Colorado Plateau are probably relict features, since lowered water tables and desiccating climatic conditions have in all likelihood resulted in reduced groundwater flow to the valley floors today. During wetter climatic episodes of the Quaternary (from about 2.6

million years ago to the present), which probably coincided with periods of mountain glaciation, spring sapping activity would have been more pronounced. Under modern climatic conditions, the results of past spring-sapping processes are obscured by the modifying action of non-sapping morphogenetic processes.

Influence of Structure

The role of structure in drainage development may be passive, in which case the composition of rocks and various rock discontinuities (joints, faults, and bedding) dictate the details of erosion. In this way, structure provides the boundary conditions for landscape degradation. During tectonism, such as faulting and folding, structural controls change with time, and the erosional system must adjust to changing resistances. Different structural controls also are encountered as incision of streams exposes lower units in the Earth's crust.

Over the years, Russian and eastern European investigators have emphasized structural control in geomorphic analysis. I.P. Gerasimov defined structural units of the landscape called morphostructures as terrain types generated by a combination of tectonic activity and climate. Various morphostructures are produced by alternating periods of uplift (with resulting dissection) and stabilization (yielding planation surfaces). The history of a morphostructure and regional tectonism can be studied by analyses of river terraces, planation surfaces, and correlative sedimentary deposits.

Drainage Patterns

The pattern of fluvial dissection of a landscape is of considerable importance in understanding the structural influence on drainage evolution. Dendritic patterns, so called because of their similarity to branching organic forms, are most common where rocks or sediments are flat-lying and preferential zones of structural weakness are minimal. The conveyance properties of a dendritic network are analogous to blood circulation systems and tree branching. Rectangular and angular patterns occur where faults, joints, and other linear structures introduce a grain to drainage. Where a broad tilt or regional slope occurs on a surface of otherwise uniform resistance, a parallel pattern occurs. Special drainage patterns characterize belts of parallel folds (trellis pattern), domes or volcanoes (radial pattern), and other landscape types.

In a series of tilted sediments the differential erosion of softer units, such as clay and shale, results in valleys developed perpendicular to the dip or tilt of the units. These strike valleys are paralleled by ridges of the tilted sediments called cuestas. Another term for a strike stream, which parallels the structural grain, is a longitudinal stream. In contrast, transverse streams cut across structural trends. Streams flowing down the tilted sediments of the cuesta are called dip streams because they parallel the structural dip of the strata. Streams draining the cuesta scarp into longitudinal valleys flowing opposite to the structural dip are called antidip streams.

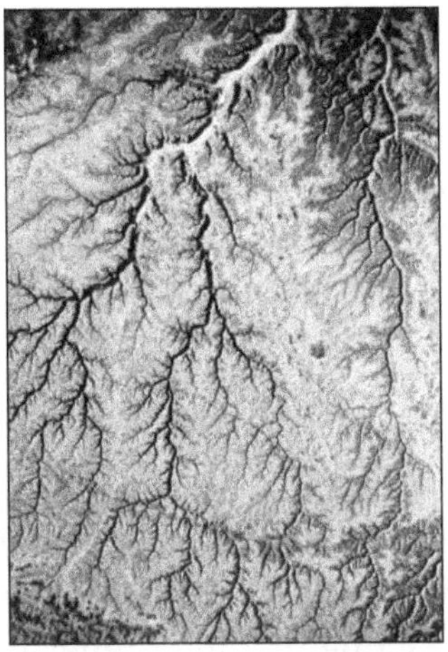

Dendritic drainage pattern developed on flat-lying limestone in central Yemen.

Cross-axial Drainage

One of the most interesting anomalies that occurs in drainage evolution is the development of stream courses across the axes of structural zones (e.g., upwarps and fold belts). Some examples of cross-axial, or discordant, drainage include rivers that appear to take the most difficult routes possible through folded regions such as the Appalachian Mountains of the United States and the Zagros Mountains of Iran. The classical studies of cross-axial drainage were made during the exploration of the Colorado River system in the 19th century by the American geologist John Wesley Powell. The Colorado River and its tributaries cross great structural upwarps. Rather than flowing around domes or plunging folds, the rivers carved canyons into what appears to be paths of greatest resistance. One theory posed by Powell for such relationships is that of antecedence. According to this view, the rivers were already in their present positions when the various anticlinal folds and upwarps began to grow. A relevant analogy is a saw into which a log is being pushed. The saw represents the river and its continuing degradation, and the log represents the growing upwarp.

Another possible origin of cross-axial drainage is superimposition. According to this theory, a cover of sedimentary material must bury older structures. The river develops on this overlying sedimentary cover and subsequently imposes its pattern across the underlying structures as they are exposed by continuing degradation.

A third explanation for cross-axial drainage is that of inheritance. In this hypothesis, an erosion surface is developed across the structure zone by long-continued planation. When the streams incise, abandoning the former planation surface, they become imposed across the structures. Alternatively, the stream may actually exploit zones of weakness or minimal resistance as it downcuts from former levels. Stream capture (also called stream piracy) occurs as more aggressively eroding portions of the drainage cut through divides. In many cases, a complex combination of the above processes probably occurs to yield the final result.

Stream capture is especially common where longitudinal streams flowing on the weaker rocks of a fold belt erode into the valleys of transverse streams that must cross the resistant strata. Sections of valley abandoned after such captures are known as wind gaps. These contrast with the water gaps that still contain transverse streams. The famous water gaps of the Appalachians are excellent examples of such patterns.

Influence of Climate

The importance of climate in landscape evolution, particularly valley development, has been emphasized by many European geomorphologists. Jean Tricart and André Cailleux of France and Julius Büdel of Germany developed climatic geomorphology as a synthesis of relief-forming processes. Climatic geomorphologists define systematic morphoclimatic zones on the globe in which relief-forming mechanisms differ as a function of climate. Some of the important morphoclimatic zones are briefly outlined in the following sections.

Periglacial Zone

The term periglacial relates to cold-climate processes and landforms. The most important periglacial influence on valleys is frost action, which produces abundant debris by freeze-thaw action on rock and soil. During the coldest periods of the Quaternary (about the last 2.6 million years), the periglacial zone was enlarged to approximately twice its present extent. Hillslopes became mantled with frost-shattered rubble that moved downslope during cycles of freezing and thawing. The relicts of this periglacial activity characterize much of the modern humid-temperate zone—e.g., in portions of Pennsylvania and Wisconsin in the United States and England and Poland.

Arid Zone

In arid regions moisture conditions are inadequate to support abundant vegetative cover of the land surface. As a result, the land is subjected to intense fluvial, eolian, and mass-wasting processes. The importance of fluvial action may seem ironic for an arid region. Although most arid regions receive little rainfall, the amount that falls is especially effective. Rare but intense arid-region rainstorms act upon a landscape that is unprotected by vegetation. Of course, some hyperarid regions receive such infrequent rainfall that fluvial processes are indeed ineffective. Nevertheless, even the most arid places on Earth show evidence of fluvial activity, either because of wetter conditions in the past or because of very rare rainstorms.

Tropical Zone

Tropical regions are dominated by dense vegetative cover and deep weathering profiles. In continuously humid tropical zones, fluvial activity is facilitated by intense rainfall but inhibited by the protective effect of rainforests. The lateritic soils of these regions, however, do not promote deep root penetration, and the vegetative cover may be undermined by fluvial erosion or mass movement. Fluvial activity may be quite intense in the tropics, especially in tectonically active areas. In more stable cratons, however, the landscape is dominated by low-relief planation surfaces. Rivers flowing on the deeply weathered regolith of these surfaces have low stream power and transport mainly fine-grained weathering products. Thus, immense contrasts in fluvial activity exist in the tropics.

At higher tropical latitudes, the continuously humid zone of the Equator changes to a zone of seasonal rainfall. Such regions have savanna vegetation because of prolonged dry seasons. Erosion rates may be extremely high in savanna environments.

Role of Climatic Change

Because the Earth's climate has changed profoundly during the Paleogene, Neogene, and Quaternary (roughly the past 65.5 million years), many landscapes are palimpsests–i.e., they are composed of relict elements produced under the influence of past climates and modern elements produced in the present climatic regime. The study of such landscape changes is sometimes called climato-genetic geomorphology. Some researchers in the field, notably Büdel, have maintained that little of the extant relief in humid temperate regions of the Earth results from modern relief-forming processes. Rather, they believe, much of the familiar humid temperate landscape is inherited from past climatic conditions, including periglacial, arid, and tropical.

Büdel focused attention on differences in the nature of valley formation as a function of climate through the history of a landscape. He argued that very rapid valley formation accompanied periods of periglacial activity in central Europe. Modern rivers in the region seem less effective at valley incision. Most of them flow on fills within great bedrock valleys, indicating that aggradation rather than downcutting is occurring today. Prior to the Quaternary phase of valley cutting, central Europe in the early Cenozoic seems to have experienced a prolonged period of planation, resulting in low-relief plains. Büdel proposed that the remnants of these Cenozoic plains, now preserved as broad planar uplands, are inherited from a time of tropical planation. Remnants of a former tropical regolith on the uplands provide some evidence for this hypothesis.

Fluvial Landscape

Large waterfalls are comparatively rare the world over, since the stream gradation process drains lakes and removes falls and rapids. But when tectonic activity fractures and dislocates crustal blocks, erosion may be hard pressed to keep up. As a result, we see great waterfalls on several large rivers in the African Rift Valley region. One of these is Victoria Falls, on the Zambezi River, shown in figure.

Another class of large waterfalls involves new river channels resulting from glacial activity in the Ice Age. Large moving ice sheets eroded and deposited sediment, creating lakes and shifting river courses in northern continental regions. Niagara Falls is a prime example. The outlet from Lake Erie into Lake Ontario, the Niagara River, runs over a gently inclined layer of limestone, beneath which lies easily eroded shale. The river has gradually eroded the edge of the limestone layer, producing a steep gorge marked by Niagara Falls at its head. The Niagara Power project uses the falls' 52-m (171-ft) drop to generate hydroelectric power. Water is withdrawn upstream from the falls and is carried in tunnels to generating plants located about 6 km (4 mi) downstream from the falls.

Located on the Zambezi River at the border of Zimbabve and Zambia in southern Africa, Victoria Falls is one of the world'scenic wonders.

Aggradation and Alluvial Terraces

Graded streams are delicately adjusted to their supply of water and rock waste from upstream sources. So they are highly sensitive to any changes in those inputs. Changes in climate or vegetation cover affect the discharge and stream load at downstream points, and these changes in turn require channel readjustments. One such change is the buildup of alluvium in valley floors.

Picture a stream transporting sediment. What happens if the bed load increases beyond the transporting capacity of the stream? The coarse sediment will start to accumulate along the section of channel where the excess load was introduced. These deposits of sand, gravel, and pebbles will raise the elevation of the streambed, a process called aggradation. As more bed materials accumulate, the stream channel gradient steepens and the flow speeds up. This velocity increase makes it easier for the stream to drag bed materials downstream, spreading them over the channel floor at more and more distant downstream sections. In this way, sediment introduced at the head of a stream will gradually spread along the whole length of the stream.

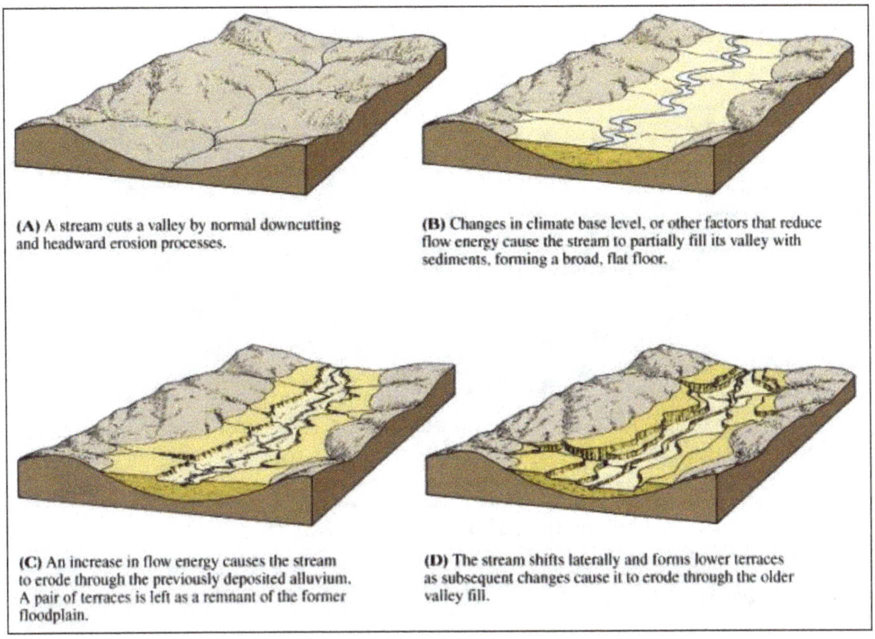

(A) A stream cuts a valley by normal downcutting and headward erosion processes.

(B) Changes in climate base level, or other factors that reduce flow energy cause the stream to partially fill its valley with sediments, forming a broad, flat floor.

(C) An increase in flow energy causes the stream to erode through the previously deposited alluvium. A pair of terraces is left as a remnant of the former floodplain.

(D) The stream shifts laterally and forms lower terraces as subsequent changes cause it to erode through the older valley fill.

Alluvial terrace formation.

Alluvial terraces attract human settlement because–unlike valley-bottom floodplains–they aren't subject to annual flooding, and they are easier to cultivate than hillslopes, which can be steep and rocky. Terraces are easily tilled and make prime agricultural land. Towns, roads, and railroads are also easily laid out on the flat ground of a terrace.

Alluvial Rivers and their Floodplains

As time passes, the floodplain is widened, so that broad areas of floodplain lie on both sides of the river channel. Civil engineers have given the name alluvial river to a large river of very low channel gradient. It flows on a thick floodplain of alluvium constructed by the river itself in earlier stages of its activity. An alluvial river in a humid environment normally experiences overbank floods each year or two.

Figure illustrates the typical landforms of an alluvial river and its floodplain, including bluffs, ox-bow lakes, natural levees, and backswamps. Bluffs are cut when meanders excavate older terraces or preexisting materials. Ox-bow lakes remain when meanders are cut off. When the floodplain is inundated, sand and silt are deposited in a zone next to the channel, creating natural levees. These flanks of high ground along the channel are normally the sites of towns and villages. Between the levees and the bluffs is lower ground, called the backswamp, which is subject to frequent flooding. It is best suited to agriculture, as regular overbank flooding of the backswamp provides a new layer of silt and infuses the soil with dissolved minerals. The resupply of nutrients helps floodplain soils retain their remarkable fertility, even though they may be located in regions of rainfall surplus from which these nutrients are normally leached away.

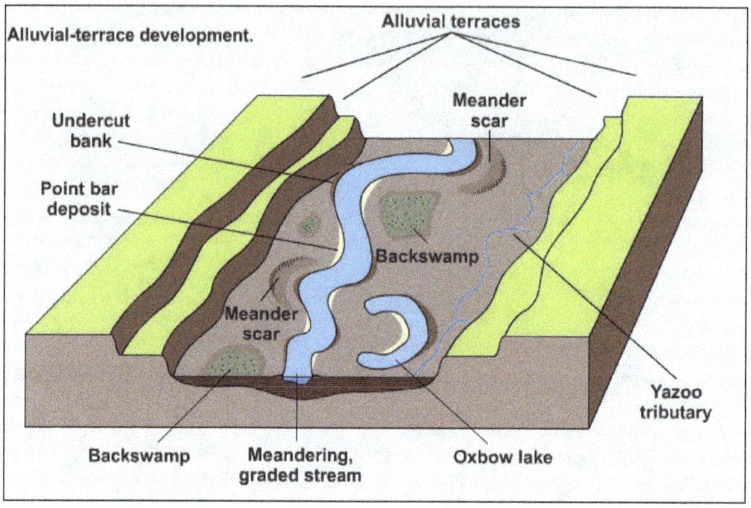

Floodplain landforms of an alluvial river.

Entrenched Meanders

What happens when a broadly meandering river is uplifted by rapid tectonic activity? The uplift increases the river's gradient, and in turn, its velocity, so that it cuts downward into the bedrock below. This forms a steep-walled inner gorge. On either side of the gorge there will be the former floodplain, now a flat terrace high above river level. Any river deposits left on the terrace are rapidly stripped off by runoff because floods no longer reach the terraces to restore eroded sediment.

The meanders become impressed into the bedrock, passing on their meandering pattern to the inner gorge. We call these sinuous bends entrenched meanders to distinguish them from the floodplain meanders of an alluvial river. Although entrenched meanders are not free to shift about as floodplain meanders do, they can slowly enlarge, producing cutoffs that can leave a high, round hill separated from the valley wall by the deep abandoned river channel and the shortened river course. Such hills formed ideal natural fortifications. Many European fortresses of the Middle Ages were built on such cutoff meander spurs. Occasionally, if the bedrock includes a strong, massive sandstone formation, the meander cutoff can leave a natural bridge formed by the narrow meander neck.

Fluvial Processes in an Arid Climate

Desert regions look strikingly different from humid regions in both vegetation and landforms. Obviously, the lower precipitation makes the difference. Vegetation is sparse or absent, and land surfaces are mantled with mineral material—sand, gravel, rock fragments, or bedrock itself.

Although deserts have low precipitation, rain falls in dry climates as well as in moist, and most landforms of desert regions are formed by running water. A particular locality in a dry desert may experience heavy rain only once in several years. But when rain does fall, stream channels carry water and perform important work as agents of erosion, transportation, and deposition. Fluvial processes are especially effective in shaping desert landforms because of the sparse vegetation cover.

The few small plants that survive offer little or no protection to soil or bedrock. Without a thick vegetative cover to protect the ground and hold back the swift downslope flow of water, large quantities of coarse rock debris are swept into the streams. A dry channel is transformed in a few minutes into a flash flood of muddy water heavily charged with rock fragments. Water enters and leaves a stream channel differently in arid climates than in humid climates. In a humid region, the high water table slopes toward a stream channel, and ground water seeps into the streambed, producing permanent (perennial) streams. In arid regions, the water table lies far below the channel floor. Where a stream flows across a plain of gravel and sand, water is lost from the channel by seepage. As a result, aggradation occurs and braided channels are common. Streams of desert regions are often short and end in alluvial deposits or on the floors of shallow, dry lakes.

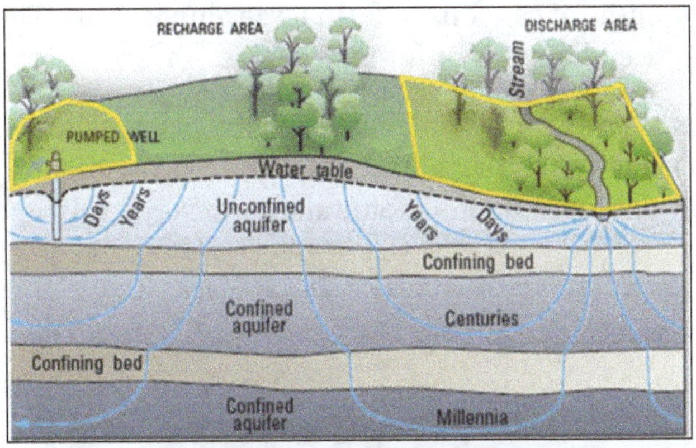

Stream water flow to ground water.

Alluvial Fans

One very common landform built by braided, aggrading streams is the alluvial fan. This is a low cone of alluvial sands and gravels resembling an open fan. The central point of the fan lies at the mouth of a canyon or ravine. The fan is built out on an adjacent plain. Alluvial fans are of many sizes. Some desert fans are many kilometers across.

Fans are built by streams carrying heavy loads of coarse rock waste from a mountain or an upland region. The braided channel shifts constantly, but its position is firmly fixed at the canyon mouth. The lower part of the channel, below the apex, sweeps back and forth—accounting for the fan form and the downward slope in all radial directions away from the apex. Alluvial fans are primary sites of ground water reservoirs in the south-western United States. In many fan areas, we have rapidly lowered the water table by pumping for irrigation. It will take an extremely long time to recharge these ground water reserves from precipitation. One serious side effect of removing too much ground water is subsidence.

Landscape of Mountainous Deserts

Where tectonic activity has recently produced block faulting in an area of continental desert, the assemblage of fluvial landforms is particularly diverse and interesting. The basin-and-range region of the western United States, which includes large parts of Nevada and Utah, south-eastern California, southern Arizona, and New Mexico, is an example.

Mountainous deserts: The initial uplift creates two uplifted fault blocks with a down-dropped block between them. Although denudation acts on the uplifted blocks as they are being raised, we have shown them as very little modified at the time tectonic activity has ceased. At first, the faces of the fault block are extremely steep. They are scored with deep ravines, and talus blocks form cones at the bases of the blocks.

At a later stage of erosion, streams have carved the mountain blocks into a rugged landscape of deep canyons and high divides. Rock waste from these steep mountain slopes is carried from the mouths of canyons to form large alluvial fans. The fan deposits form a continuous apron extending far out into the basins.

In the center of the desert basin is a dry lakebed, or playa, where fine sediments and precipitated salts produce a flat basin floor. In some playas, shallow water forms a salt lake. Figure is an air photograph of a mountainous desert landscape showing many of these features.

Landscape of mountainous deserts.

A gently sloping rock surface, thinly veneered with alluvium, sometimes stretches from the mountain flank toward the playa. This surface—a pediment—may be an ancient low-angle fault plane; an old surface created by many years of weathering during a wetter climate; or a surface formed by retreat of the mountain front.

Geographic Cycle

The Earth's fluvial landscapes are quite diverse. They range from mountain regions of steep slopes and rugged peaks to regions of gentle hills and valleys, to nearly flat plains that stretch from horizon to horizon. We can think of these constantly changing landscapes as stages of evolution in a cycle that begins with rapid uplift by plate tectonic activity and follows with long erosion by streams in a graded condition. This cycle, called the geographic cycle, was first described by William Morris Davis, a prominent geographer and geomorphologist of the late nineteenth and early twentieth centuries.

In Davis's model, the landscape begins at a youthful stage of steep slopes and river channels. In the mature stage, hills are rounded and slopes are gentler. In old age, the landscape is an undulating surface drained by slow and sluggish streams. With rapid uplift, rejuvenation occurs and the cycle begins anew.

Anabranch

Anabranches at the junction of the Yukon River and the Koyukuk River in Alaska.

An anabranch is a section of a river or stream that diverts from the main channel or stem of the watercourse and rejoins the main stem downstream. Local anabranches can be the result of small islands in the watercourse. In larger anabranches, the flow can diverge for a distance of several kilometers before rejoining the main channel.

The term anabranch, in its hydrological meaning, is used more frequently in Australia than in the rest of the English-speaking world.

The term anabranching river describes a river with many anabranches, whilst an anastomosing river is an organic-rich subset of this river type.

The term braided river describes watercourses which are divided by small islands into multiple channel threads within a single main channel, but the term does not describe the multiple channels of an anabranching river.

A distributary is a branch of a river that does not rejoin the main channel; these are common on and near river deltas. A bayou is often an anabranch. An anabranch that gets cut off from the main channel becomes an oxbow lake.

Examples of Anabranch:

- The Magdalena River, in Colombia, bifurcates in El Banco in two branches: Brazo de Loba and Brazo de Mompox. The first one, Brazo de Loba is the main navigable channel and is 177 km long. Brazo de Mompox is thus the anabranch. There are several other branches that join those two main branches, interconnected by swamps, the main ones being Quitasol, Chicagua and La Victoria. Between Brazo de Loba and Brazo de Mompox is the island of Mompox or Margarita Island, a sunken geological block of 2.200 square km between the faults of Chicagua and El Romeral.
- In the Fraser River delta of British Columbia, Canada, North Arm Fraser River, Middle Arm Fraser River, and South Arm Fraser River each fall into Georgia Strait separately. On the other hand, Annacis Island splits (South Arm) Fraser River into the (main) Annieville Channel and the (smaller) Annacis Channel, which rejoin below the island.
- In western New South Wales the Great Darling Anabranch is the ancestral path of the Darling River, dividing south of Menindee and extending for 460 km before joining the Murray River. The anabranch contains flowing water only in wetter years.
- In the Lower Simpson Desert, in South Australia, Kallakoopah Creek is an anabranch of the Warburton River that flows during major Warburton floods of over 5 m in depth. The Kallakoopah is approximately 400 km long and members of the Lake Eyre Yacht Club and others have used it to sail a circular route to Lake Eyre by traveling down the Kallakoopah and returning via the Warburton.
- The Bahr el Zeraf in southern Sudan splits from the Bahr al Jabal section of the White Nile and flows for 240 km (150 mi), before rejoining the White Nile proper upriver from Malakal.
- Approximately 15 km out of Charters Towers (Queensland/Australia) is a section of the Burdekin River known as 'the anabranch'. This strip of water is separate from the main flood way, and is typically dry during the sunnier months. Once there is consistent rain, it will meet up with the river again and flood with fresh water. It is well known for barramundi fishing.
- Probably because waterways played an important role in infrastructure, anabranches and the resulting river islands are often found in old cities:
    - The Île de la cité in Paris.
    - Manhattan, in New York.

Meander

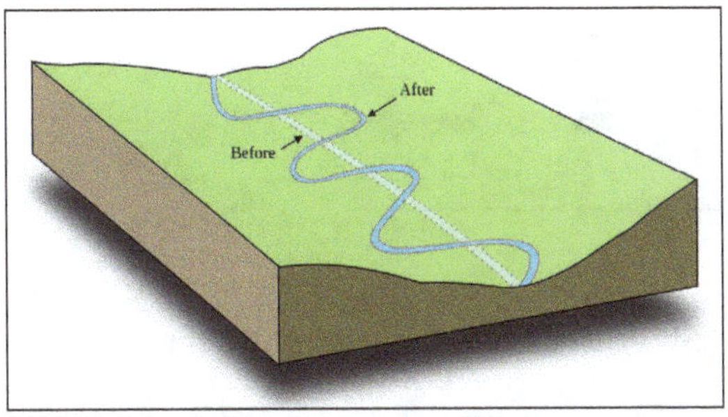

A stream bed following a tilted valley. The maximum gradient is along the down-valley axis represented by a hypothetical straight channel. Meanders develop, which lengthen the course of the stream, decreasing the gradient.

Meanders of the Rio Cauto at Guamo Embarcadero, Cuba.

A meander is one of a series of regular sinuous curves, bends, loops, turns, or windings in the channel of a river, stream, or other watercourse. It is produced by a stream or river swinging from side to side as it flows across its floodplain or shifts its channel within a valley. A meander is produced by a stream or river as it erodes the sediments comprising an outer, concave bank (cut bank) and deposits this and other sediment downstream on an inner, convex bank which is typically a point bar. The result of sediments being eroded from the outside concave bank and their deposition on an inside convex bank is the formation of a sinuous course as a channel migrates back and forth across the down-valley axis of a floodplain. The zone within which a meandering stream shifts its channel across either its floodplain or valley floor from time to time is known as a meander belt. It typically ranges from 15 to 18 times the width of the channel. Over time, meanders migrate downstream, sometimes in such a short time as to create civil engineering problems for local municipalities attempting to maintain stable roads and bridges.

The degree of meandering of the channel of a river, stream, or other watercourse is measured by its sinuosity. The sinuosity of a watercourse is the ratio of the length of the channel to the straight line down-valley distance. Streams or rivers with a single channel and sinuosities of 1.5 or more are defined as meandering streams or rivers.

Governing Physics

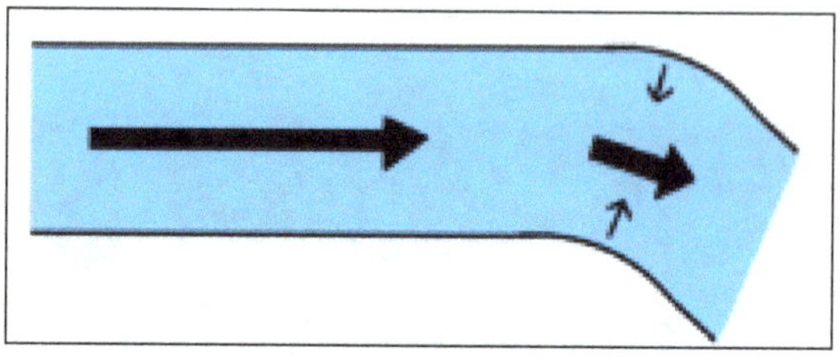

Straight channel culminating in a single bend.

When a fluid is introduced to an initially straight channel which then bends, the sidewalls induce a pressure gradient that causes the fluid to alter course and follow the bend. From here, two opposing processes occur: (1) irrotational flow and (2) secondary flow. For a river to meander, secondary flow must dominate.

Irrotational flow: From Bernoulli's equations, high pressure results in low velocity. Therefore, in the absence of secondary flow we would expect low fluid velocity at the outside bend and high fluid velocity at the inside bend. This classic fluid mechanics result is irrotational vortex flow. In the context of meandering rivers, its effects are dominated by those of secondary flow.

Secondary flow: A force balance exists between pressure forces pointing to the inside bend of the river and centrifugal forces pointing to the outside bend of the river. In the context of meandering rivers, a boundary layer exists within the thin layer of fluid that interacts with the river bed. Inside that layer and following standard boundary-layer theory, the velocity of the fluid is effectively zero. Centrifugal force, which depends on velocity, is also therefore effectively zero. Pressure force, however, remains unaffected by the boundary layer. Therefore, within the boundary layer, pressure force dominates and fluid moves along the bottom of the river from the outside bend to the inside bend. This initiates helicoidal flow: Along the river bed, fluid roughly follows the curve of the channel but is also forced toward the inside bend; away from the river bed, fluid also roughly follows the curve of the channel but is forced, to some extent, from the inside to the outside bend.

The higher velocities at the outside bend lead to higher shear stresses and therefore result in erosion. Similarly, lower velocities at the inside bend cause lower sheer stresses and deposition occurs. Thus meander bends erode at the outside bend, causing the river to becoming increasingly sinuous (until cutoff events occur). Deposition at the inside bend occurs such that for most natural meandering rivers, the river width remains nearly constant, even as the river evolves.

Even where the river is not forced to bend by a natural obstacle, Coriolis force of the earth can cause a small imbalance in velocity distribution such that velocity on one bank is higher than on the other. This can trigger erosion on one bank and deposition of sediment on the other.

Meander Geometry

Uvac canyon meander, Serbia.

Meanders on the River Clyde, Scotland.

The technical description of a meandering watercourse is termed meander geometry or meander planform geometry. It is characterized as an irregular waveform. Ideal waveforms, such as a sine wave, are one line thick, but in the case of a stream the width must be taken into consideration. The bankfull width is the distance across the bed at an average cross-section at the full-stream level, typically estimated by the line of lowest vegetation.

As a waveform the meandering stream follows the down-valley axis, a straight line fitted to the curve such that the sum of all the amplitudes measured from it is zero. This axis represents the overall direction of the stream.

At any cross-section the flow is following the sinuous axis, the centerline of the bed. Two consecutive crossing points of sinuous and down-valley axes define a meander loop. The meander is two consecutive loops pointing in opposite transverse directions. The distance of one meander along the down-valley axis is the meander length or wavelength. The maximum distance from the down-valley axis to the sinuous axis of a loop is the meander width or amplitude. The course at that point is the apex.

In contrast to sine waves, the loops of a meandering stream are more nearly circular. The curvature varies from a maximum at the apex to zero at a crossing point (straight line), also called an inflection, because the curvature changes direction in that vicinity. The radius of the loop is the straight line perpendicular to the down-valley axis intersecting the sinuous axis at the apex. As the loop is not ideal, additional information is needed to characterize it. The orientation angle is the angle between sinuous axis and down-valley axis at any point on the sinuous axis.

Concave bank and convex bank, Great Ouse Relief Channel, England.

A loop at the apex has an outer or concave bank and an inner or convex bank. The meander belt is defined by an average meander width measured from outer bank to outer bank instead of from centerline to centerline. If there is a flood plain, it extends beyond the meander belt. The meander is then said to be free—it can be found anywhere in the flood plain. If there is no flood plain, the meanders are fixed.

Various mathematical formulae relate the variables of the meander geometry. As it turns out some numerical parameters can be established, which appear in the formulae. The waveform depends ultimately on the characteristics of the flow but the parameters are independent of it and apparently are caused by geologic factors. In general the meander length is 10–14 times, with an average 11 times, the fullbank channel width and 3 to 5 times, with an average of 4.7 times, the radius of curvature at the apex. This radius is 2–3 times the channel width.

Meander of the River Cuckmere in Southern England.

A meander has a depth pattern as well. The cross-overs are marked by riffles, or shallow beds, while at the apices are pools. In a pool direction of flow is downward, scouring the bed material. The major volume, however, flows more slowly on the inside of the bend where, due to decreased velocity, it deposits sediment.

The line of maximum depth, or channel, is the thalweg or thalweg line. It is typically designated the borderline when rivers are used as political borders. The thalweg hugs the outer banks and returns to center over the riffles. The meander arc length is the distance along the thalweg over one meander. The river length is the length along the centerline.

Formation

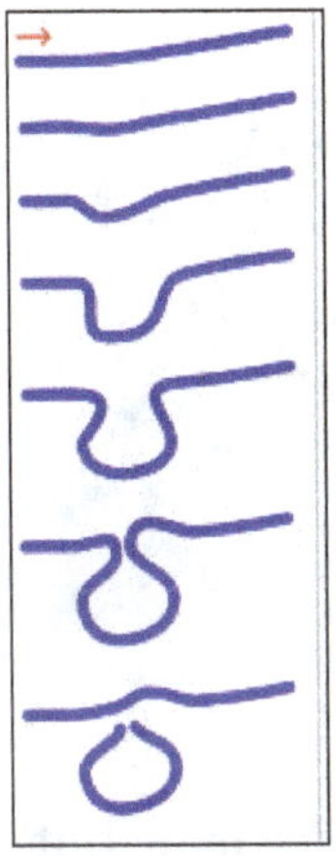

Life history of a meander.

Meander formation is a result of natural factors and processes. The waveform configuration of a stream is constantly changing. Fluid flows around a bend in a vortex. Once a channel begins to follow a sinusoidal path, the amplitude and concavity of the loops increase dramatically due to the effect of helical flow sweeping dense eroded material towards the inside of the bend, and leaving the outside of the bend unprotected and therefore vulnerable to accelerated erosion, forming a positive feedback loop. In the words of Elizabeth A. Wood:

> 'This process of making meanders seems to be a self-intensifying process in which greater curvature results in more erosion of the bank, which results in greater curvature'.

The cross-current along the floor of the channel is part of the secondary flow and sweeps dense eroded material towards the inside of the bend. The cross-current then rises to the surface near the inside and flows towards the outside, forming the helical flow. The greater the curvature of the bend, and the faster the flow, the stronger is the cross-current and the sweeping.

Due to the conservation of angular momentum the speed on the inside of the bend is faster than on the outside.

Since the flow velocity is diminished, so is the centrifugal pressure. The pressure of the super-elevated column prevails, developing an unbalanced gradient that moves water back across the bottom from the outside to the inside. The flow is supplied by a counter-flow across the surface from the inside to the outside. This entire situation is very similar to the Tea leaf paradox. This secondary flow carries sediment from the outside of the bend to the inside making the river more meandering.

As to why streams of any size become sinuous in the first place, there are a number of theories, not necessarily mutually exclusive.

Stochastic Theory

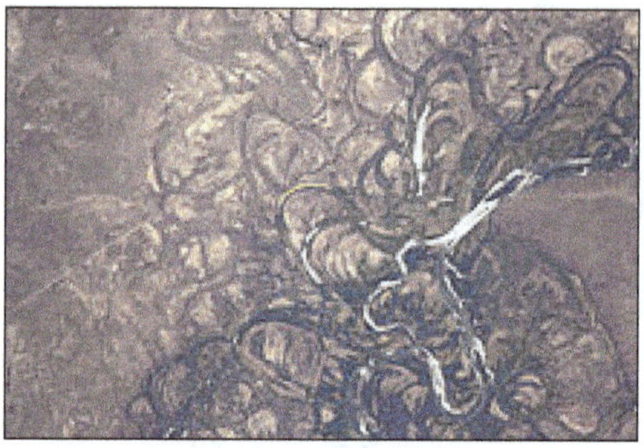

Meander scars, oxbow lakes and abandoned meanders in the broad flood plain of the Rio Negro, Argentina.

The stochastic theory can take many forms but one of the most general statements is that of Scheidegger: 'The meander train is assumed to be the result of the stochastic fluctuations of the direction of flow due to the random presence of direction-changing obstacles in the river path.' Given a flat, smooth, tilted artificial surface, rainfall runs off it in sheets, but even in that case adhesion of water to the surface and cohesion of drops produce rivulets at random. Natural surfaces are rough and erodible to different degrees. The result of all the physical factors acting at random is channels that are not straight, which then progressively become sinuous. Even channels that appear straight have a sinuous thalweg that leads eventually to a sinuous channel.

Equilibrium Theory

In the equilibrium theory, meanders decrease the stream gradient until equilibrium between the erodibility of the terrain and the transport capacity of the stream is reached. A mass of water descending must give up potential energy, which, given the same velocity at the end of the drop as at the beginning, is removed by interaction with the material of the stream bed. The shortest distance; that is, a straight channel, results in the highest energy per unit of length, disrupting the banks more, creating more sediment and aggrading the stream. The presence of meanders allows the stream to adjust the length to an equilibrium energy per unit length in which the stream carries away all the sediment that it produces.

Geomorphic and Morphotectonic Theory

Geomorphic refers to the surface structure of the terrain. Morphotectonic means having to do with the deeper, or tectonic (plate) structure of the rock. The features included under these categories are not random and guide streams into non-random paths. They are predictable obstacles that instigate meander formation by deflecting the stream. For example, the stream might be guided into a fault line (morphotectonic).

Associated Landforms

Cut Bank

A cut bank is an often vertical bank or cliff that forms where the outside, concave bank of a meander cuts into the floodplain or valley wall of a river or stream. A cutbank is also known either as a river-cut cliff, river cliff, or a bluff and spelled as cutbank. Erosion that forms a cut bank occurs at the outside bank of a meander because helicoidal flow of water keeps the bank washed clean of loose sand, silt, and sediment and subjects it to constant erosion. As a result, the meander erodes and migrates in the direction of the outside bend, forming the cut bank.

As the cut bank is undermined by erosion, it commonly collapses as slumps into the river channel. The slumped sediment, having been broken up by slumping, is readily eroded and carried toward the middle of the channel. The sediment eroded from a cut bank tends to be deposited on the point bar of the next downstream meander, and not on the point bar opposite it. This can be seen in areas where trees grow on the banks of rivers; on the inside of meanders, trees, such as willows, are often far from the bank, whilst on the outside of the bend, the tree roots are often exposed and undercut, eventually leading the trees to fall into the river.

Meander Cutoff

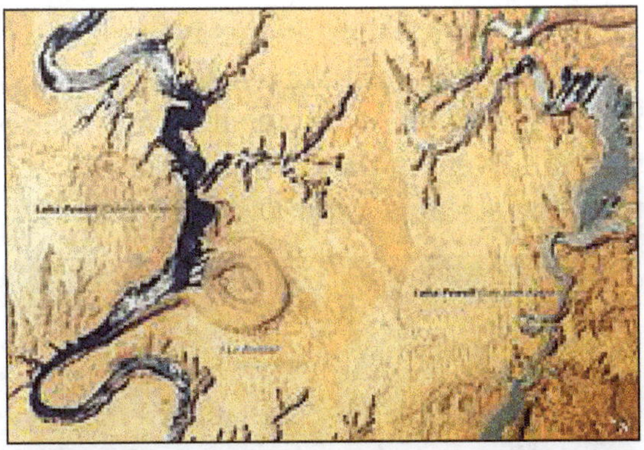

The Rincon on Lake Powell in southern Utah. It is an incised cutoff (abandoned) meander.

A meander cutoff, also known as either a cutoff meander or abandoned meander, is a meander that has been abandoned by its stream after the formation of a neck cutoff. A lake that occupies a cutoff meander is known as an oxbow lake. Cutoff meanders that have cut downward into the underlying bedrock are known in general as incised cutoff meanders. As in the case of the Anderson Bottom Rincon, incised meanders that have either steep-sided, often vertical walls, are often, but not always, known as rincons in the southwest United States. Rincon in English is a nontechnical word in the southwest United States for either a small secluded valley, an alcove or angular recess in a cliff, or a bend in a river.

Incised Meanders

Glen Canyon, US

The meanders of a stream or river that has cut its bed down into the bedrock are known as either incised, intrenched, entrenched, inclosed or ingrown meanders. Some Earth scientists recognize and use a finer subdivision of incised meanders. Thornbury argues that incised or inclosed meanders are synonyms that are appropriate to describe any meander incised downward into bedrock and defines enclosed or entrenched meanders as a subtype of incised meanders (inclosed meanders) characterized by a symmetrical valley sides. He argues that the symmetrical valley sides are the direct result of rapid down-cutting of a watercourse into bedrock. In addition, as proposed by Rich, Thornbury argues that incised valleys with a pronounced asymmetry of cross section, which he called ingrown meanders, are the result of the lateral migration and incision of a meander during a period of slower channel downcutting. Regardless, the formation of both entrenched meanders and ingrown meanders is thought to require that base level falls as a result of either relative change in mean sea level, isostatic or tectonic uplift, the breach of an ice or landslide dam, or regional tilting. Classic examples of incised meanders are associated with rivers in the Colorado Plateau, the Kentucky River Palisades in central Kentucky, and streams in the Ozark Plateau.

Goosenecks of the San Juan River, SE Utah. There is a cut-off meander at right center.

It was initially either argued or presumed that an incised meander is characteristic of an antecedent stream or river that had incised its channel into underlying strata. An antecedent stream or river is one that maintains its original course and pattern during incision despite the changes in underlying rock topography and rock types. However, later geologists argue that the shape of an incised meander is not always, if ever, "inherited," e.g., strictly from an antecedent meandering stream where it meander pattern could freely develop on a level floodplain. Instead, they argue that as fluvial incision of bedrock proceeds, the stream course is significantly modified by variations in rock type and fractures, faults, and other geological structures into either lithologically conditioned meanders or structurally controlled meanders.

Oxbow Lakes

The oxbow lake, which is the most common type of fluvial lake, is a crescent-shaped lake that derives its name from its distinctive curved shape. Oxbow lakes are also known as cutoff lakes. Such lakes form regularly in undisturbed floodplains as a result of the normal process of fluvial meandering. Either a river or stream forms a sinuous channel as the outer side of its bends are eroded away and sediments accumulate on the inner side, which forms a meandering horseshoe-shaped bend. Eventually as the result of its meandering, the fluvial channel cuts through the narrow neck of the meander and forms a cutoff meander. The final break-through of the neck, which is called a neck cutoff, often occurs during a major flood because that is when the watercourse is out of its banks and can flow directly across the neck and erode it with the full force of the flood.

After a cutoff meander is formed, river water flows into its end from the river builds small delta-like feature into either end of it during floods. These delta-like features block either end of the cutoff meander to form a stagnant oxbow lake that is separated from the flow of the fluvial channel and independent of the river. During floods, the flood waters deposit fine-grained sediment into the oxbow lake. As a result, oxbow lakes tend to become filled in with fine-grained, organic-rich sediments over time.

Point Bar

A point bar, which is also known as a meander bar, is a fluvial bar that is formed by the slow, often episodic, addition of individual accretions of noncohesive sediment on the inside bank of a meander by the accompanying migration of the channel toward its outer bank. This process is called lateral accretion. Lateral accretion occurs mostly during high water or floods when the point bar is submerged. Typically, the sediment consists of either sand, gravel, or a combination of both. The sediment comprising some point bars might grade downstream into silty sediments. Because of the decreasing velocity and strength of current from the thalweg of the channel to the upper surface of point bar when the sediment is deposited the vertical sequence of sediments comprising a point bar becomes finer upward within an individual point bar. For example, it is typical for point bars to fine upward from gravel at the base to fine sands at the top. The source of the sediment is typically upstream cut banks from which sand, rocks and debris has been eroded, swept, and rolled across the bed of the river and downstream to the inside bank of a river bend. On the inside bend, this sediment and debris is eventually deposited on the slip-off slope of a point bar.

Scroll-bars

Scroll-bars are a result of continuous lateral migration of a meander loop that creates an asymmetrical ridge and swale topography on the inside of the bends. The topography is generally parallel to the meander, and is related to migrating bar forms and back bar chutes, which carve sediment from the outside of the curve and deposit sediment in the slower flowing water on the inside of the loop, in a process called lateral accretion. Scroll-bar sediments are characterized by cross-bedding and a pattern of fining upward. These characteristics are a result of the dynamic river system, where larger grains are transported during high energy flood events and then gradually die down, depositing smaller material with time. Deposits for meandering rivers are generally homogeneous and laterally extensive unlike the more heterogeneous braided river deposits. There are two distinct patterns of scroll-bar depositions; the eddy accretion scroll bar pattern and the point-bar scroll pattern. When looking down the river valley they can be distinguished because the point-bar scroll patterns are convex and the eddy accretion scroll bar patterns are concave.

Scroll bars often look lighter at the tops of the ridges and darker in the swales. This is because the tops can be shaped by wind, either adding fine grains or by keeping the area unvegetated, while the darkness in the swales can be attributed to silts and clays washing in during high water periods. This added sediment in addition to water that catches in the swales is in turn is a favorable environment for vegetation that will also accumulate in the swales.

Slip-off Slope

Depending upon whether a meander is part of an entrenched river or part of a freely meandering river within a floodplain, the term slip-off slope can refer to two different fluvial landforms that comprise the inner, convex, bank of a meander loop. In case of a freely meandering river on a floodplain, a slip-off slope is the inside, gently sloping bank of a meander on which sediments episodically accumulate to form a point bar as a river meanders. This type of slip-off slope is located opposite the cutbank. This term can also be applied to the inside, sloping bank of a meandering tidal channel.

In case of an entrenched river, a slip-off slope is a gently sloping bedrock surface that rises from the inside, concave bank of an asymmetrically entrenched river. This type of slip-off slope is often covered by a thin, discontinuous layer of alluvium. It is produced by the gradual outward migration of the meander as a river cuts downward into bedrock. A terrace on the slip-off slope of a meander spur, known as slip-off slope terrace, can formed by a brief halt during the irregular incision by an actively meandering river.

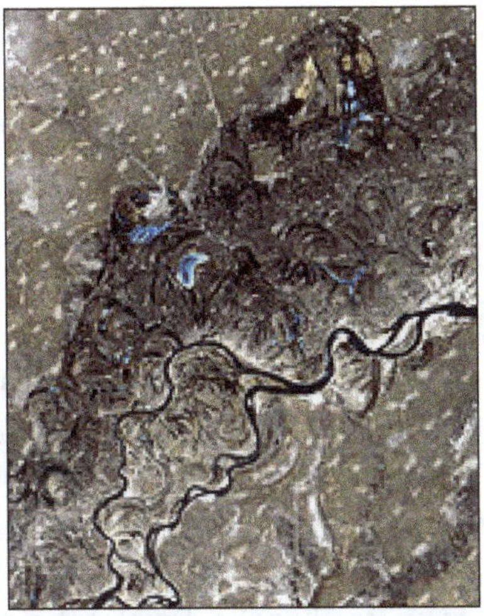

Meanders, scroll-bars and oxbow lakes in the Songhua River.

The meander ratio or sinuosity index is a means of quantifying how much a river or stream meanders (how much its course deviates from the shortest possible path). It is calculated as the length of the stream divided by the length of the valley. A perfectly straight river would have a meander ratio of 1 (it would be the same length as its valley), while the higher this ratio is above 1, the more the river meanders.

Sinuosity indices are calculated from the map or from an aerial photograph measured over a distance called the reach, which should be at least 20 times the average fullbank channel width. The length of the stream is measured by channel, or thalweg, length over the reach, while the bottom value of the ratio is the downvalley length or air distance of the stream between two points on it defining the reach.

The sinuosity index plays a part in mathematical descriptions of streams. The index may require elaboration, because the valley may meander as well—i.e., the downvalley length is not identical to the reach. In that case the valley index is the meander ratio of the valley while the channel index is the meander ratio of the channel. The channel sinuosity index is the channel length divided by the valley length and the standard sinuosity index is the channel index divided by the valley index. Distinctions may become even more subtle.

Sinuosity Index has a non-mathematical utility as well. Streams can be placed in categories arranged by it; for example, when the index is between 1 and 1.5 the river is sinuous, but if between 1.5 and 4, then meandering. The index is a measure also of stream velocity and sediment load, those quantities being maximized at an index of 1 (straight).

Meander Cutoff

A meander cutoff, the natural form of a cutting or cut in a river occurs when a pronounced meander (hook) in a river is breached by a flow that connects the two closest parts of the hook to form a new channel, a full loop. The steeper drop in gradient (slope) causes the river flow gradually to abandon the meander which will silt up with sediment from deposition. Cutoffs are a natural part of the evolution of a meandering river.

Meandering rivers flow higher and hence with more total flow, pressure and erosion on the outside of their bends due to forming a vortex as in a stirred coffee cup and consequently the river erodes more the outer bank. On the inside bend of a river, the level is lower, secondary flow moves sand and gravel across the river bed creating shallows and point bars, and friction of air and perturbances of the bed act against a higher proportion of the column of water, being shorter, slowing the water to varying degrees. Rivers are commonly described and interpreted by their sinuosity. The term is equally used to describe the actual incidence of and potential tendency of a river to curve or meander over its length. It is expressed as the ratio of the distance between two distant points in a river following the middle-of-the-river course of the river as compared with the straight distance between those points. Three conventional categorizations of rivers or their reaches exist. Meandering rivers a sinuosity value/ratio of greater than 1.5. A sinuosity value of less than 1.1 is a "straight" river. Between these values, a river is described as sinuous which describes those in a transitory state between the two states. Braided rivers do not follow this same convention. Meandering rivers trend in the direction of increasing sinuosity.

Cutoff Channel

A river constantly evolves and as it does, meanders that were once a part of the river are abandoned in favor of a route that is more efficient for a river to take. As these old meanders are cutoff from the rest of the river, a new channel, or cutoff channel, is formed.

- Neck cutoff: A river bend intersects itself. On the outer bends of a river, where water is flowing fastest, the river will erode away the land between adjacent outer bends and cause them to become close to each other; this leads to their intersection.
- Chute cutoff: A channel cuts directly across the land bypassing an entire meandering loop in the river and abandoning it.

Formation

A chute cutoff channel can form during a flood resulting in an overbank flow where water goes over the banks of the river creating erosion of the surrounding landscape. More studies need to be done on how the magnitude of these floods and their recurrence interval are related to how often these chute cutoff channels form. Neck cutoff channels are commonly formed the same way when an overbank flow occurs during a flood and the narrow piece of land between a bend in a meander is eroded away; this is known as rush-cutting. A meander can also be cutoff by a channel due to excess sediment upstream as a result of high erosion rates. This leads to a cutoff channel forming since a river might no longer be able to carry that sediment through the bend efficiently, so the river forms a new path for it to flow. Meander cutoffs can also be formed by humans; by removing a beaver dam, the likelihood that meander cutoff channels will be formed downstream increases. A cutoff channel can be engineered for the purpose of navigation, traditionally for water mill leats and for controlling the possibility of any future flood where done on the lower reaches away from the tide. These meander cutoffs straighten a river. Many rivers are transformed by humans, becoming less sinuous.

Oxbow Lake

When either of these meander cutoff processes takes place a bend of the river is left behind forming, in many instances, an oxbow lake. An oxbow lake forms after there has been deposition of sediment, by the new cutoff channel flowing adjacent to it, at the entrances of the abandoned bend; this seals the bend off from the rest of the river. Oxbow lakes have been shown to be an important habitat for various species of wildlife. Recent efforts have been made to protect these important bodies of water from harmful practices such as agricultural use. One proposed method to restore these oxbow lakes has been dredging. Dredging will remove sediment from the lake's floor and will increase the lake's depth. Collins Lake in Scotia, New York is an example of this method. Oxbow lakes can be valuable for recreational purposes and in Salix, Iowa, Browns Lake's water level was increased for recreational use.

View along the former Mississippi River riverbed at the Tennessee/Arkansas state line near Reverie, Tennessee.

On 7 March 1876 a cutoff formed suddenly across the neck of a meander, known as the "Devil's Elbow", in the Mississippi River near Reverie, Tennessee, shortening the river's course and leaving the town connected to Arkansas, but across the new river channel from the rest of Tennessee.

Some research has been done to show importance. Cutoffs have been shown to limit the age of a river meander and thus how large that meander can get, without which in areas of extreme sinuosity and low gradients means very long, slightly slowed, sections of rivers intensifying local flood risk. Meander cutoffs influence the formation of a river's floodplain and continue to do so as the river evolves. Cutoffs can affect the way that other river bends adjacent evolve over time, increasing their height of flooding and the direct momentum of the water which can then create further cutoffs and affect other bends further downstream throughout the river. Meander cutoffs directly reduce and tend to indirectly reduce a river's sinuosity, thus straightening out a river's channel. Understanding the processes that form meander cutoffs can allow one to predict how a river will evolve in the future which is important for agricultural businesses and controlling future floods.

Glacial and Glaciofluvial Landscapes

Glacial landform is any product of flowing ice and meltwater such landforms are being produced today in glaciated areas, such as Greenland, Antarctica, and many of the world's higher mountain ranges. In addition, large expansions of present-day glaciers have recurred during the course of Earth history. At the maximum of the last ice age, which ended about 20,000 to 15,000 years ago, more than 30 percent of the Earth's land surface was covered by ice. Consequently, if they have not been obliterated by other landscape-modifying processes since that time, glacial landforms may still exist in regions that were once glaciated but are now devoid of glaciers.

Eskers, or narrow ridges of gravel and sand left by a retreating glacier, wind through western Nunavut, Canada.

Periglacial features, which form independently of glaciers, are nonetheless a product of the same cold climate that favours the development of glaciers, and so are treated here as well.

Types of Glaciers

There are numerous types of glaciers, but it is sufficient here to focus on two broad classes: Mountain, or valley, glaciers and continental glaciers, or ice sheets, (including ice caps).

Generally, ice sheets are larger than valley glaciers. The main difference between the two classes, however, is their relationship to the underlying topography. Valley glaciers are rivers of ice usually found in mountainous regions, and their flow patterns are controlled by the high relief in those areas. In map view, many large valley glacier systems, which have numerous tributary glaciers that join to form a large "trunk glacier," resemble the roots of a plant. Pancake like ice sheets, on the other hand, are continuous over extensive areas and completely bury the underlying landscape beneath hundreds or thousands of metres of ice. Within continental ice sheets, the flow is directed more or less from the centre outward. At the periphery, however, where ice sheets are much thinner, they may be controlled by any substantial relief existing in the area. In this case, their borders may be lobate on a scale of a few kilometres, with tongue like protrusions called outlet glaciers. Viewed by themselves, these are nearly indistinguishable from the lower reaches of a large valley glacier system. Consequently, many of the landforms produced by valley glaciers and continental ice sheets are similar or virtually identical, though they often differ in magnitude.

Glacial Erosion

Two processes, internal deformation and basal sliding, are responsible for the movement of glaciers under the influence of gravity. The temperature of glacier ice is a critical condition that affects these processes. For this reason, glaciers are classified into two main types, temperate and polar, according to their temperature regime. Temperate glaciers are also called isothermal glaciers, because they exist at the pressure-melting point (the melting temperature of ice at a given pressure) throughout their mass. The ice in polar, or cold glaciers, in contrast, is below the pressure-melting point. Some glaciers have an intermediate thermal character. For example, subpolar glaciers are temperate in their interior parts, but their margins are cold-based. This classification is a broad generalization, however, because the thermal condition of a glacier may show wide variations in both space and time.

Internal deformation, or strain, in glacier ice is a response to shear stresses arising from the weight of the ice (ice thickness) and the degree of slope of the glacier surface. Internal deformation occurs by movement within and between individual ice crystals (slow creep) and by brittle failure (fracture), which arises when the mass of ice cannot adjust its shape rapidly enough by the creep process to take up the stresses affecting it. The relative importance of these two processes is greatly influenced by the temperature of the ice. Thus, fractures due to brittle failure under tension, known as crevasses, are usually much deeper in polar ice than they are in temperate ice.

The temperature of the basal ice is an important influence upon a glacier's ability to erode its bed. When basal temperatures are below the pressure-melting point, the ability of the ice mass to slide on the bed (basal sliding) is inhibited by the adhesion of the basal ice to the frozen bed beneath. Basal sliding is also diminished by the greater rigidity of polar ice: this reduces the rate of creep, which, in turn, reduces the ability of the more rigid ice to deform around obstacles on the glacier bed. Thus, the flow of cold-based glaciers is predominantly controlled by internal deformation, with proportionately low rates of basal sliding. For this reason, rates of abrasion are commonly low beneath polar glaciers, and slow rates of erosion commonly result. Equally, the volume of meltwater is frequently very low, so that the extent of sediments and landforms derived from polar glaciers is limited.

Temperate glaciers, being at the pressure-meeting point, move by both mechanisms, with basal sliding being the more important. It is this sliding that enables temperate glaciers to erode their beds and carve landforms so effectively. Ice is, however, much softer and has a much lower shear strength than most rocks, and pure ice alone is not capable of substantially eroding anything other than unconsolidated sediments. Most temperate glaciers have a basal debris zone from several centimetres to a few metres thick that contains varying amounts of rock debris in transit. In this respect, glaciers act rather like sheets of sandpaper; while the paper itself is too soft to sand wood, the adherent hard grains make it a powerful abrasive system. The analogy ends here, however, for the rock debris found in glaciers is of widely varying sizes—from the finest rock particles to large boulders—and also generally of varied types as it includes the different rocks that a glacier is overriding. For this reason, a glacially abraded surface usually bears many different "tool-marks," from microscopic scratches to gouges centimetres deep and tens of metres long. Over thousands of years glaciers may erode their substrate to a depth of several tens of metres by this mechanism, producing a variety of streamlined landforms typical of glaciated landscapes.

Several other processes of glacial erosion are generally included under the terms glacial plucking or quarrying. This process involves the removal of larger pieces of rock from the glacier bed. Various explanations for this phenomenon have been proposed. Some of the mechanisms suggested are based on differential stresses in the rock caused by ice being forced to flow around bedrock obstacles. High stress gradients are particularly important, and the resultant tensile stresses can pull the rock apart along pre-existing joints or crack systems. These pressures have been shown to be sufficient to fracture solid rock, thus making it available for removal by the ice flowing above it. Other possibilities include the forcing apart of rock by the pressure of crystallization produced beneath the glacier as water derived from the ice refreezes (regelation) or because of temperature fluctuations in cavities under the glacier. Still another possible mechanism involves hydraulic pressures of flowing water known to be present, at least temporarily, under nearly all temperate glaciers. It is hard to determine which process is dominant because access to the base of active glaciers is rarely possible. Nonetheless, investigators know that larger pieces of rock are plucked from the glacier bed and contribute to the number of abrasive “tools” available to the glacier at its base. Other sources for the rock debris in glacier ice may include rockfalls from steep slopes bordering a glacier or unconsolidated sediments overridden as a glacier advances.

Glacial Deposition

Debris in the glacial environment may be deposited directly by the ice (till) or, after reworking, by meltwater streams (outwash). The resulting deposits are termed glacial drift.

As the ice in a valley glacier moves from the area of accumulation to that of ablation, it acts like a conveyor belt, transporting debris located beneath, within, and above the glacier toward its terminus or, in the case of an ice sheet, toward the outer margin. Near the glacier margin where the ice velocity decreases greatly is the zone of deposition. As the ice melts away, the debris that was originally frozen into the ice commonly forms a rocky and muddy blanket over the glacier margin. This layer often slides off the ice in the form of mudflows. The resulting deposit is called a flow-till by some authors. On the other hand, the debris may be laid down more or less in place as the ice melts away around and beneath it. Such deposits are referred to as melt-out till, and sometimes as ablation till. In many cases, the material located between a moving glacier and its bedrock bed is severely sheared, compressed, and “over-compacted.” This type of deposit is called lodgment till. By definition, till is any material laid down directly or reworked by a glacier. Typically, it is a mixture of rock fragments and boulders in a fine-grained sandy or muddy matrix (non-stratified drift). The exact composition of any particular till, however, depends on the materials available to the glacier at the time of deposition. Thus, some tills are made entirely of lake clays deformed by an overriding glacier. Other tills are composed of river gravels and sands that have been “bulldozed” and striated during a glacial advance. Tills often contain some of the tools that glaciers use to abrade their bed. These rocks and boulders bear striations, grooves, and facets, and characteristic till-stones are commonly shaped like bullets or flat-irons. Till-boulders of a rock type different from the bedrock on which they are deposited are dubbed “erratics.” In some cases, erratics with distinctive lithologies can be traced back to their source, enabling investigators to ascertain the direction of ice movement of ice sheets in areas where striations either are absent or are covered by till or vegetation.

Meltwater deposits, also called glacial outwash, are formed in channels directly beneath the glacier or in lakes and streams in front of its margin. In contrast to till, outwash is generally bedded or laminated (stratified drift), and the individual layers are relatively well sorted according to grain size. In most cases, gravels and boulders in outwash are rounded and do not bear striations or grooves on their surfaces, since these tend to wear off rapidly during stream transport. The grain size of individual deposits depends not only on the availability of different sizes of debris but also on the velocity of the depositing current and the distance from the head of the stream. Larger boulders are deposited by rapidly flowing creeks and rivers close to the glacier margin. Grain size of deposited material decreases with increasing distance from the glacier. The finest fractions, such as clay and silt, may be deposited in glacial lakes or ponds or transported all the way to the ocean.

Finally, it must be stressed that most glacier margins are constantly changing chaotic masses of ice, water, mud, and rocks. Ice-marginal deposits thus are of a highly variable nature over short distances, as is much the case with till and outwash as well.

Erosional Landforms

Small-scale Features of Glacial Erosion

Glacial erosion is caused by two different processes: abrasion and plucking. Nearly all glacially scoured erosional landforms bear the tool-marks of glacial abrasion provided that they have not been removed by subsequent weathering. Even though these marks are not large enough to be called landforms, they constitute an integral part of any glacial landscape and thus warrant description here. The type of mark produced on a surface during glacial erosion depends on the size and shape of the tool, the pressure being applied to it, and the relative hardnesses of the tool and the substrate.

Rock Polish

The finest abrasive available to a glacier is the so-called rock flour produced by the constant grinding at the base of the ice. Rock flour acts like jewelers' rouge and produces microscopic scratches, which with time smooth and polish rock surfaces, often to a high lustre.

Striations

These are scratches visible to the naked eye, ranging in size from fractions of a millimetre to a few millimetres deep and a few millimetres to centimetres long. Large striations produced by a single tool may be several centimetres deep and wide and tens of metres long.

Because the striation-cutting tool was dragged across the rock surface by the ice, the long axis of a striation indicates the direction of ice movement in the immediate vicinity of that striation. Determination of the regional direction of movement of former ice sheets, however, requires measuring hundreds of striation directions over an extended area because ice moving close to the base of a glacier is often locally deflected by bedrock obstacles. Even when such a regional study is conducted, additional information is frequently needed in low-relief areas to determine which end of the striations points down-ice toward the former outer margin of the glacier. On an outcrop scale, such information can be gathered by studying "chatter marks." These crescentic gouges and lunate fractures are caused by the glacier dragging a rock or boulder over a hard and brittle rock surface and forming a series of sickle-shaped gouges. Such depressions in the bedrock are steep-sided on their "up-glacier" face and have a lower slope on their down-ice side. Depending on whether the horns of the sickles point up the glacier or down it, the chatter marks are designated crescentic gouges or lunate fractures. Another small-scale feature that allows absolute determination of the direction in which the ice moved is what is termed knob-and-tail. A knob-and-tail is formed during glacial abrasion of rocks that locally contain spots more resistant than the surrounding rock, as is the case, for example, with silicified fossils in limestone. After abrasion has been active for some time, the harder parts of the rock form protruding knobs as the softer rock is preferentially eroded away around them. During further erosion, these protrusions protect the softer rock on their lee side and a tail forms there, pointing from the knob to the margin of the glacier. The scale of these features depends primarily on the size of the in-homogeneities in the rock and ranges from fractions of millimetres to metres.

P-forms and Glacial Grooves

These features, which extend several to tens of metres in length, are of uncertain origin. P-forms (P for plastically molded) are smooth-walled, linear depressions which may be straight, curved, or sometimes hairpin-shaped and measure tens of centimetres to metres in width and depth. Their cross sections are often semicircular to parabolic, and their walls are commonly striated parallel to their long axis, indicating that ice once flowed in them. Straight P-forms are frequently called glacial grooves, even though the term is also applied to large striations, which, unlike the P-forms, were cut by a single tool. Some researchers believe that P-forms were not carved directly by the ice but rather were eroded by pressurized mud slurries flowing beneath the glacier.

Erosional Landforms of Valley Glaciers

Many of the world's higher mountain ranges—e.g., the Alps, the North and South American Cordilleras, the Himalayas, and the Southern Alps in New Zealand, as well as the mountains of Norway, including those of Spitsbergen—are partly glaciated today. During periods of the Pleistocene, such glaciers were greatly enlarged and filled most of the valleys with ice, even reaching far beyond the mountain front in certain places. Most scenic alpine landscapes featuring sharp mountain peaks, steep-sided valleys, and innumerable lakes and waterfalls are a product of several periods of glaciation.

Erosion is generally greater than deposition in the upper reaches of a valley glacier, whereas deposition exceeds erosion closer to the terminus. Accordingly, erosional landforms dominate the landscape in the high areas of glaciated mountain ranges.

Cirques, Tarns, U-shaped Valleys, Arêtes and Horns

The heads of most glacial valleys are occupied by one or several cirques (or corries). A cirque is an amphitheatre-shaped hollow with the open end facing down-valley. The back is formed by an arcuate cliff called the headwall. In an ideal cirque, the headwall is semicircular in plan view. This situation, however, is generally found only in cirques cut into flat plateaus. More common are headwalls angular in map view due to irregularities in height along their perimeter. The bottom of many cirques is a shallow basin, which may contain a lake. This basin and the base of the adjoining headwall usually show signs of extensive glacial abrasion and plucking. Even though the exact process of cirque formation is not entirely understood, it seems that the part of the headwall above the glacier retreats by frost was shattering and ice wedging. The rock debris then falls either onto the surface of the glacier or into the randkluft or bergschrund. Both names describe the crevasse between the ice at the head of the glacier and the cirque headwall. The rocks on the surface of the glacier are successively buried by snow and incorporated into the ice of the glacier. Because of a downward velocity component in the ice in the accumulation zone, the rocks are eventually moved to the base of the glacier. At that point, these rocks, in addition to the rock debris from the bergschrund, become the tools with which the glacier erodes, striates, and polishes the base of the headwall and the bottom of the cirque.

During the initial growth and final retreat of a valley glacier, the ice often does not extend beyond the cirque. Such a cirque glacier is probably the main cause for the formation of the basin scoured into the bedrock bottom of many cirques. Sometimes these basins are "over-deepened" several tens of metres and contain lakes called tarns.

In contrast to the situation in a stream valley, all debris falling or sliding off the sides and the headwalls of a glaciated valley is immediately removed by the flowing ice. Moreover, glaciers are generally in contact with a much larger percentage of a valley's cross section than equivalent rivers or creeks. Thus glaciers tend to erode the bases of the valley walls to a much greater extent than do streams, whereas a stream erodes an extremely narrow line along the lowest part of a valley. The slope of the adjacent valley walls depends on the stability of the bedrock and the angle of repose of the weathered rock debris accumulating at the base of and on the valley walls. For this reason, rivers tend to form V-shaped valleys. Glaciers, which inherit V-shaped stream valleys, reshape them drastically by first removing all loose debris along the base of the valley walls and then preferentially eroding the bedrock along the base and lower sidewalls of the valley. In this way, glaciated valleys assume a characteristic parabolic or U-shaped cross profile, with relatively wide and flat bottoms and steep, even vertical sidewalls. By the same process, glaciers tend to narrow the bedrock divides between the upper reaches of neighbouring parallel valleys to jagged, knife-edge ridges known as arêtes. Arêtes also form between two cirques facing in opposite directions. The low spot, or saddle, in the arête between two cirques is called a col. A higher mountain often has three or more cirques arranged in a radial pattern on its flanks. Headward erosion of these cirques finally leaves only a sharp peak flanked by nearly vertical headwall cliffs, which are separated by arêtes. Such glacially eroded mountains are termed horns, the most widely known of which is the Matterhorn in the Swiss Alps.

Hanging Valleys

Large valley glacier systems consist of numerous cirques and smaller valley glaciers that feed ice into a large trunk glacier. Because of its greater ice discharge, the trunk glacier has greater erosive capability in its middle and lower reaches than smaller tributary glaciers that join it there. The main valley is therefore eroded more rapidly than the side valleys. With time, the bottom of the main valley becomes lower than the elevation of the tributary valleys. When the ice has retreated, the tributary valleys are left joining the main valley at elevations substantially higher than its bottom. Tributary valleys with such unequal or discordant junctions are called hanging valleys. In extreme cases where a tributary joins the main valley high up in the steep part of the U-shaped trough wall, waterfalls may form after deglaciation, as in Yosemite and Yellowstone national parks in the western United States.

Paternoster Lakes

Some glacial valleys have an irregular, longitudinal bedrock profile, with alternating short, steep steps and longer, relatively flat portions. Even though attempts have been made to explain this feature in terms of some inherent characteristic of glacial flow, it seems more likely that differential erodibility of the underlying bedrock is the real cause of the phenomenon. Thus the steps are probably formed by harder or less fractured bedrock, whereas the flatter portions between the steps are underlain by softer or more fractured rocks. In some cases, these softer areas have been excavated by a glacier to form shallow bedrock basins. If several of these basins are occupied by lakes along one glacial trough in a pattern similar to beads on a string, they are called paternoster lakes by analogy with a string of rosary beads.

Roches Moutonnées

These structures are bedrock knobs or hills that have a gently inclined, glacially abraded, and streamlined stoss side (i.e., one that faces the direction from which the overriding glacier impinged) and a steep, glacially plucked lee side. They are generally found where jointing or fracturing in the bedrock allows the glacier to pluck the lee side of the obstacle. In plain view, their long axes are often, but not always, aligned with the general direction of ice movement.

Rock Drumlins

A feature similar to roches moutonnées, rock drumlins are bedrock knobs or hills completely streamlined, usually with steep stoss sides and gently sloping lee sides. Both roches moutonnées and rock drumlins range in length from several metres to several kilometres and in height from tens of centimetres to hundreds of metres. They are typical of both valley and continental glaciers. The larger ones, however, are restricted to areas of continental glaciation.

Erosional Landforms of Continental Glaciers

In contrast to valley glaciers, which form exclusively in areas of high altitude and relief, continental glaciers, including the great ice sheets of the past, occur in high and middle latitudes in both hemispheres, covering landscapes that range from high alpine mountains to low-lying areas with negligible relief. Therefore, the landforms produced by continental glaciers are more diverse and widespread. Yet, just like valley glaciers, they have an area where erosion is the dominant process and an area close to their margins where net deposition generally occurs. The capacity of a continental glacier to erode its substrate has been a subject of intense debate. All of the areas formerly covered by ice sheets show evidence of are ally extensive glacial scouring. The average depth of glacial erosion during the Pleistocene probably did not exceed a few tens of metres, however. This is much less than the deepening of glacial valleys during mountain glaciation. One of the reasons for the apparent limited erosional capacity of continental ice sheets in areas of low relief may be the scarcity of tools available to them in these regions. Rocks cannot fall onto a continental ice sheet in the accumulation zone, because the entire landscape is buried. Thus, all tools must be quarried by the glacier from the underlying bedrock. With time, this task becomes increasingly difficult as bedrock obstacles are abraded and streamlined. Nonetheless, the figure for depth of glacial erosion during the Pleistocene cited above is an average value, and locally several hundreds of metres of bedrock were apparently removed by the great ice sheets. Such enhanced erosion seems concentrated at points where the glaciers flowed from hard, resistant bedrock onto softer rocks or where glacial flow was channelized into outlet glaciers.

As a continental glacier expands, it strips the underlying landscape of the soil and debris accumulated at the preglacial surface as a result of weathering. The freshly exposed harder bedrock is then eroded by abrasion and plucking. During this process, bedrock obstacles are shaped into streamlined "whaleback" forms, such as roches moutonnées and rock drumlins. The adjoining valleys are scoured into rock-floored basins with the tools plucked from the lee sides of roches moutonnées. The long axes of the hills and valleys are often preferentially oriented in the direction of ice flow. An area totally composed of smooth whaleback forms and basins is called a streamlined landscape.

Streams cannot erode deep basins because water cannot flow uphill. Glaciers, on the other hand, can flow uphill over obstacles at their base as long as there is a sufficient slope on the upper ice surface pointing in that particular direction. Therefore the great majority of the innumerable lake basins and small depressions in formerly glaciated areas can only be a result of glacial erosion. Many of these lakes, such as the Finger Lakes in the U.S. state of New York, are aligned parallel to the direction of regional ice flow. Other basins seem to be controlled by preglacial drainage systems. Yet, other depressions follow the structure of the bedrock, having been preferentially scoured out of areas underlain by softer or more fractured rock.

A number of the largest freshwater lake basins in the world (e.g., the Great Lakes or the Great Slave Lake and Great Bear Lake in Canada) are situated along the margins of the Precambrian shield of North America. Many researchers believe that glacial erosion was especially effective at these locations because the glaciers could easily abrade the relatively soft sedimentary rocks to the south with hard, resistant crystalline rocks brought from the shield areas that lie to the north. Nonetheless, further research is necessary to determine how much of the deepening of these features can be ascribed to glacial erosion, as opposed to other processes such as tectonic activity or pre-glacial stream erosion.

Fjords are found along some steep, high-relief coast-lines where continental glaciers formerly flowed into the sea. They are deep, narrow valleys with U-shaped cross sections that often extend inland for tens or hundreds of kilometres and are now partially drowned by the ocean. These troughs are typical of the Norwegian coast, but they also are found in Canada, Alaska, Iceland, Greenland, Antarctica, New Zealand, and southernmost Chile. The floor and steep walls of fjords show ample evidence of glacial erosion. The long profile of many fjords, including alternating basins and steps, is very similar to that of glaciated valleys. Toward the mouth, fjords may reach great depths, as in the case of Sogn Fjord in southern Norway where the maximum water depth exceeds 1,300 metres. At the mouth of a fjord, however, the floor rises steeply to create a rock threshold, and water depths decrease markedly. At Sogn Fjord the water at this "threshold" is only 150 metres deep, and in many fjords the rock platform is covered by only a few metres of water. The exact origin of fjords is still a matter of debate. While some scientists favour a glacial origin, others believe that much of the relief of fjords is a result of tectonic activity and that glaciers only slightly modified preexisting large valleys. In order to erode Sogn Fjord to its present depth, the glacier occupying it during the maximum of the Pleistocene must have been 1,800 to 1,900 metres thick. Such an ice thickness may seem extreme, but even now, during an interglacial period, the Skelton Glacier in Antarctica has a maximum thickness of about 1,450 metres. This outlet glacier of the Antarctic ice sheet occupies a trough, which in places is more than one kilometre below sea level and would become a fjord in the event of a large glacial retreat.

Depositional Landforms of Valley Glaciers

Moraines

As a glacier moves along a valley, it picks up rock debris from the valley walls and floor, transporting it in, on, or under the ice. As this material reaches the lower parts of the glacier where ablation is dominant, it is concentrated along the glacier margins as more and more debris melts out of the ice. If the position of the glacier margin is constant for an extended amount of time, larger accumulations of glacial debris will form at the glacier margin. In addition, a great deal of material is rapidly flushed through and out of the glacier by meltwater streams flowing under, within, on, and next to the glacier. Part of this stream-load is deposited in front of the glacier close to its snout. There, it may mix with material brought by, and melting out from, the glacier as well as with material washed in from other, non-glaciated tributary valleys. If the glacier then advances or re-advances after a time of retreat, it will "bulldoze" all the loose material in front of it into a ridge of chaotic debris that closely hugs the shape of the glacier snout. Any such accumulation of till melted out directly from the glacier or piled into a ridge by the glacier is a moraine. Large valley glaciers are capable of forming moraines a few hundred metres high and many hundreds of metres wide. Linear accumulations of till formed immediately in front of or on the lower end of the glacier are end moraines.

The moraines formed along the valley slopes next to the side margins of the glacier are termed lateral moraines. During a single glaciation, a glacier may form many such moraine arcs, but all the smaller moraines, which may have been produced during standstills or short advances while the glacier moved forward to its outermost ice position, are generally destroyed as the glacier resumes its advance. The end moraine of largest extent formed by the glacier (which may not be as extensive as the largest ice advance) during a given glaciation is called the terminal moraine of that glaciation. Successively smaller moraines formed during standstills or small readvances as the glacier retreats from the terminal moraine position are recessional moraines.

Flutes

The depositional equivalent of erosional knob-and-tail structures are known as flutes. Close to the lower margin, some glaciers accumulate so much debris beneath them that they actually glide on a bed of pressurized muddy till. As basal ice flows around a pronounced bedrock knob or a boulder lodged in the substrate, a cavity often forms in the ice on the lee side of the obstacle because of the high viscosity of the ice. Any pressurized muddy paste present under the glacier may then be injected into this cavity and deposited as an elongate tail of till, or flute. The size depends mainly on the size of the obstacle and on the availability of subglacial debris. Flutes vary in height from a few centimetres to tens of metres and in length from tens of centimetres to kilometres, even though very large flutes are generally limited to continental ice sheets.

Depositional Landforms of Continental Glaciers

Many of the deposits of continental ice sheets are very similar to those of valley glaciers. Terminal, end, and recessional moraines are formed by the same process as with valley glaciers, but they can be much larger. Morainic ridges may be laterally continuous for hundreds of kilometres, hundreds of metres high, and several kilometres wide. Since each moraine forms at a discreet position of the ice margin, plots of end moraines on a map of suitable scale allow the reconstruction of ice sheets at varying stages during their retreat.

In addition to linear accumulations of glacial debris, continental glaciers often deposit a more or less continuous, thin (less than 10 metres) sheet of till over large areas, which is called ground moraine. This type of moraine generally has a "hummocky" topography of low relief, with alternating small till mounds and depressions. Swamps or lakes typically occupy the low-lying areas. Flutes are a common feature found in areas covered by ground moraine.

Another depositional landform associated with continental glaciation is the drumlin, a streamlined, elongate mound of sediment. Such structures often occur in groups of tens or hundreds, which are called drumlin fields. The long axis of individual drumlins is usually aligned parallel to the direction of regional ice flow. In long profile, the stoss side of a drumlin is steeper than the lee side. Some drumlins consist entirely of till, while others have bedrock cores draped with till. The till in many drumlins has been shown to have a "fabric" in which the long axes of the individual rocks and sand grains are aligned parallel to the ice flow over the drumlin. Even though the details of the process are not fully understood, drumlins seem to form subglacially close to the edge of an ice sheet, often directly down-ice from large lake basins overridden by the ice during an advance. The difference between a rock drumlin and a drumlin is that the former is an erosional bedrock knob, whereas the latter is a depositional till feature.

Meltwater Deposits

Much of the debris in the glacial environment of both valley and continental glaciers is transported, reworked, and laid down by water. Whereas glaciofluvial deposits are formed by meltwater streams, glaciolacustrine sediments accumulate at the margins and bottoms of glacial lakes and ponds.

Glaciofluvial Deposits

The discharge of glacial streams is highly variable, depending on the season, time of day, and cloud cover. Maximum discharges occur during the afternoon on warm, sunny summer days, and minima on cold winter mornings. Beneath or within a glacier, the water flows in tunnels and is generally pressurized during periods of high discharge. In addition to debris washed in from unglaciated highlands adjacent to the glacier, a glacial stream can pick up large amounts of debris along its path at the base of the glacier. For this reason, meltwater streams issuing forth at the snout of a valley glacier or along the margin of an ice sheet are generally laden to transporting capacity with debris. Beyond the glacier margin, the water, which is no longer confined by the walls of the ice tunnel, spreads out and loses some of its velocity. Because of the decreased velocity, the stream must deposit some of its load. As a result, the original stream channel is choked with sediments, and the stream is forced to change its course around the obstacles, often breaking up into many winding and shifting channels separated by sand and gravel bars. The highly variable nature of the sediments laid down by such a braided stream reflects the unstable environment in which they form. Lenses of fine-grained, cross-bedded sands are often interbedded laterally and vertically with stringers of coarse, bouldery gravel. Since the amount of sediment laid down generally decreases with distance from the ice margin, the deposit is often wedge-shaped in cross section, ideally gently sloping off the end moraine formed at that ice position and thinning downstream. The outwash is then said to be "graded to" that particular moraine. In map view, the shape of the deposit depends on the surrounding topography. Where the valleys are deep enough not to be buried by the glaciofluvial sediments, as in most mountainous regions, the resulting elongate, planar deposits are termed valley trains. On the other hand, in low-relief areas the deposits of several ice-marginal streams may merge to form a wide outwash plain, or sandur.

If the ice margin stabilizes at a recessional position during glacial retreat, another valley train or sandur may be formed inside of the original one. Because of the downstream thinning of the outwash at any one point in the valley, the recessional deposit will be lower than and inset into the outer, slightly older outwash plain. Flat-topped remnants of the older plain may be left along the valley sides; these are called terraces. Ideally each recessional ice margin has a terrace graded to it, and these structures can be used in addition to moraines to reconstruct the positions of ice margins through time. In some cases where the glacier either never formed moraines or where the moraines were obliterated by the outwash or postglacial erosion, terraces are the only means of ice margin reconstruction.

Streams that flow over the terminus of a glacier often deposit stratified drift in their channels and in depressions on the ice surface. As the ice melts away, this ice-contact stratified drift slumps and partially collapses to form stagnant ice deposits. Isolated mounds of bedded sands and gravels deposited in this manner are called kames. Kame terraces form in a similar manner but between the lateral margin of a glacier and the valley wall. Glacial geologists sometimes employ the term kame moraine to describe deposits of stratified drift laid down at an ice margin in the arcuate shape of a moraine. Some researchers, however, object to the use of the term moraine in this context because the deposit is not composed of till.

In some cases, streams deposit stratified drift in subglacial or englacial tunnels. As the ice melts away, these sinuous channel deposits may be left as long linear gravel ridges called eskers. Some eskers deposited by the great ice sheets of the Pleistocene can be traced for hundreds of kilometres, even though most esker segments are only a few hundred metres to kilometres long and a few to tens of metres high.

Kettles, potholes, or ice pits are steep-sided depressions typical of many glacial and glaciofluvial deposits. Kettles form when till or outwash is deposited around ice blocks that have become separated from the active glacier by ablation. Such "stagnant" ice blocks may persist insulated under a mantle of debris for hundreds of years. When they finally melt, depressions remain in their place, bordered by slumped masses of the surrounding glacial deposits. Many of the lakes in areas of glacial deposition are water-filled kettles and so are called kettle lakes. If a sandur or valley train contains many kettles, it is referred to as a pitted outwash plain.

Glaciolacustrine Deposits

Glacial and proglacial lakes are found in a variety of environments and in considerable numbers. Erosional lake basins have already been mentioned, but many lakes are formed as streams are dammed by the ice itself, by glacial deposits, or by a combination of these factors. Any lake that remains at a stable level for an extended period of time (e.g., hundreds or thousands of years) tends to form a perfectly horizontal, flat, terrace like feature along its beach. Such a bench may be formed by wave erosion of the bedrock or glacial sediments that form the margin of the lake, and it is called a wave-cut bench. On the other hand, it may be formed by deposition of sand and gravel from longshore currents along the margin of the lake, in which case it is referred to as a beach ridge. The width of these shorelines varies from a few metres to several hundred metres. As the lake level is lowered due to the opening of another outlet or down-cutting of the spillway, new, lower shorelines may be formed. Most former or existing glacial lakes (e.g., the Great Salt Lake and the Great Lakes in North America) have several such shorelines that can be used both to determine the former size and depth of now-extinct or shrunken lakes and to determine the amount of differential postglacial uplift because they are now tilted slightly from their original horizontal position.

Lago de Grey, Torres del Paine National Park, Chile.

Where a stream enters a standing body of water, it is forced to deposit its bedload. The coarser gravel and sand are laid down directly at the mouth of the stream as successive, steeply inclined foreset beds. The finer, suspended silt and clay can drift a bit farther into the lake, where they are deposited as almost flat-lying bottomset beds. As the sediment builds out farther into the lake (or ocean), the river deposits a thin veneer of sub-horizontal gravelly top-set beds over the foreset units. Because the foreset–topset complex often has the shape of a triangle with the mouth of the stream at one apex, such a body of sediment is called a delta. Many gravel and sand pits are located in deltas of former glacial lakes.

The flat-lying, fine-grained bottomset beds of many large former glacial lakes filled in and buried all of the pre-existing relief and are now exposed, forming perfectly flat lake plains. Cuts into these sediments often reveal rhythmically interbedded silts and clays. Some of these so-called rhythmites have been shown to be the result of seasonal changes in the proglacial environment. During the warmer summer months, the meltwater streams carry silt and clay into the lakes, and the silt settles out of suspension more rapidly than the clay. A thicker, silty summer layer is thus deposited. During the winter, as the surface of the lake freezes and the meltwater discharge into it ceases, the clays contained in the lake water slowly settle out of suspension to form a thin winter clay layer. Such lacustrine deposits with annual silt and clay "couplets" are known as varves.

Periglacial Landforms

In the cold, or periglacial (near-glacial), areas adjacent to and beyond the limit of glaciers, a zone of intense freeze-thaw activity produces periglacial features and landforms. This happens because of the unique behaviour of water as it changes from the liquid to the solid state. As water freezes, its volume increases about 9 percent. This is often combined with the process of differential ice growth, which traps air, resulting in an even greater increase in volume. If confined in a crack or pore space, such ice and air mixtures can exert pressures of about 200,000 kilopascals (29,000 pounds per square inch). This is enough to break the enclosing rock. Thus freezing water can be a powerful agent of physical weathering. If multiple freeze-and-thaw cycles occur, the growth of ice crystals fractures and moves material by means of frost shattering and frost heaving, respectively. In addition, in permafrost regions where the ground remains frozen all year, characteristic landforms are formed by perennial ice.

Felsenmeers, Talus and Rock Glaciers

In nature, the tensional strength of most rocks is exceeded by the pressure of water crystallizing in cracks. Thus, repeated freezing and thawing not only forms potholes in poorly constructed roads but also is capable of reducing exposed bedrock outcrops to rubble. Many high peaks are covered with frost-shattered angular rock fragments. A larger area blanketed with such debris is called a felsenmeer, from the German for "sea of rocks." The rock fragments can be transported downslope by flowing water or frost-induced surface creep, or they may fall off the cliff from which they were wedged by the ice. Accumulations of this angular debris at the base of steep slopes are known as talus. Owing to the steepness of the valley sides of many glacial troughs, talus is commonly found in formerly glaciated mountain regions. Talus cones are formed when the debris coming from above is channelized on its way to the base of the cliff in rock chutes. As the talus cones of neighbouring chutes grow over time, they may coalesce to form a composite talus apron.

In higher mountain regions, the interior of thick accumulations of talus may remain at temperatures below freezing all year. Rain or meltwater percolating into the interstices between the rocks freezes over time, filling the entire pore space. In some cases, enough ice forms to enable the entire mass of rock and ice to move downhill like a glacier. The resulting massive, lobate, mobile feature is called a rock glacier. Some rock glaciers have been shown to contain pure ice under a thick layer of talus with some interstitial ice. These features may be the final retreat stages of valley glaciers buried under talus.

Permafrost, Patterned Ground, Solifluction Deposits and Pingos

Permafrost is ground that remains perennially frozen. It covers about 20–25 percent of the Earth's land surface today. The "active layer" of soil close to the surface of permafrost regions undergoes many seasonal and daily freeze-thaw cycles. The constant change in the volume of water tends to move the coarser particles in the soil to the surface. Further frost heaving arranges the stones and rocks according to their sizes to produce patterned ground. Circular arrangements of the larger rocks are termed stone rings. When neighbouring stone rings coalesce, they form polygonal stone nets. On steeper slopes, stone rings and stone nets are often stretched into stone stripes by slow downhill motion of the soggy active layer of the permafrost. In other areas, patterned ground is formed by vertical or sub-vertical polygonal cracks, which are initiated in the soil by contraction during extremely cold winters. During the spring thaw of the active layer, water flows into these cracks, freezes, and expands. This process is repeated year after year, and the ice-filled cracks increase in size. The resulting ice wedges are often several metres deep and a few tens of centimetres wide at the top. Along the sides of ice wedges, the soil is deformed and compressed. Because of this disturbance and sediment that may be washed into the crack as the ice melts, relict patterned ground may be preserved during a period of warmer climate long after the permafrost has thawed. Today, relict patterned ground that formed during the last ice age exists more than 1,000 kilometres to the south of the present limit of permafrost.

When the active layer of permafrost moves under the influence of gravity, the process is termed gelifluction. The soft flowing layer is often folded and draped on hillsides and at the base of slopes as solifluction, or gelifluction, lobes.

In some permafrost areas, a locally abundant groundwater supply present at a relatively shallow depth may cause the exceptional growth of ice within a confined area. The sustained supply of liquid water results in the expansion of an increasingly large, lens-shaped ice body. These conical mounds, or pingos, may be several tens of metres high and hundreds of metres in diameter.

Periglacial Landscapes

Several definitions exist for the term periglacial. The earliest definitions suggested that these geomorphic environments were located at periphery of past Pleistocene glaciers. In these environments, the landscape is dominantly influenced by frost action However, frost action also influences landscapes that were not at the margin of ancient glaciers. For this reason, we should use a broader definition of this term. This definition suggests that in a periglacial environment the effects of freezing and thawing drastically modify the ground surface. Types of modification include the displacement of soil materials, migration of groundwater, and the formation of unique landforms. More than a third of the Earth's terrestrial surface can be included in this definition.

Permafrost

Permafrost is a condition where a layer of soil, sediment, or rock below the ground surface remains frozen for a period greater than a year. Permafrost is not a necessary condition for creating periglacial landforms. However, many periglacial regions are underlain by permafrost and it influences geomorphic processes acting in this region of the world.

Permafrost is found in about 25 percent of the Earth's non-glaciated land surface. It occurs wherever the mean annual ground temperature is less than 0° Celsius. The southern limit in Canada generally correlates with an annual surface air temperature of -8° Celsius. In some areas, permafrost can be up to 1500 meters deep. Common depths of permafrost are several hundred meters. Most deposits of permafrost have an upper active layer that is between 1 to 3 meters thick. This active layer is subject to a cyclic thaw during the summer season.

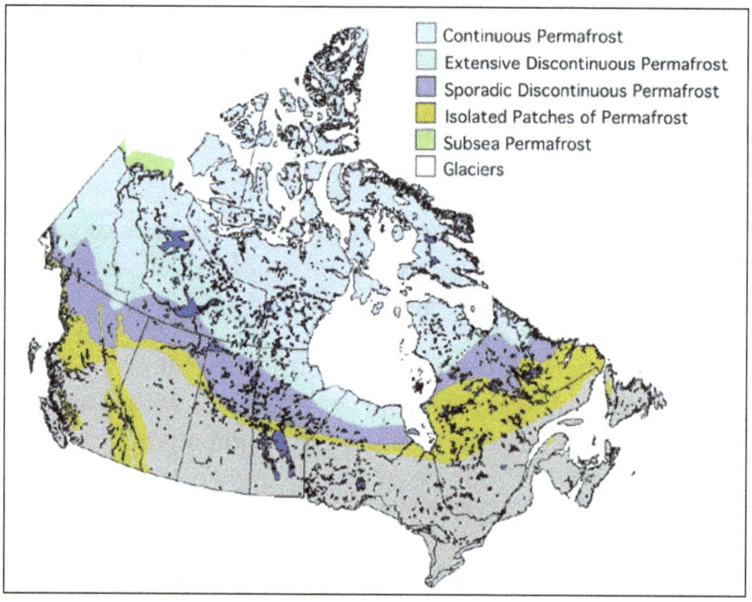

Distribution of the various types of permafrost in Canada. More than 50% of Canada is covered by some form of permafrost. This map does not identify areas of alpine permafrost.

In some places, localized unfrozen layers or taliks are located on top, underneath, or within masses of permafrost. Often in continuous permafrost areas, taliks are found under lakes because of the ability of water to store and vertically transfer heat energy. The vertical extent of the taliks found under lakes is related to the depth and volume of the overlying water body. Larger water bodies can store and transfer more heat energy downward. A closed talik is unfrozen ground that is found within a mass of permafrost. Closed taliks can develop when lakes fill in with sediment and become bogs. With the removal of the open water, summer solar radiation is now being received by a surface with a lower specific heat and poor vertical heat transfer. As a result, soil near the surface begins to freeze solid encasing a zone of unfrozen soil in permafrost. Closed taliks can also form because of groundwater flow.

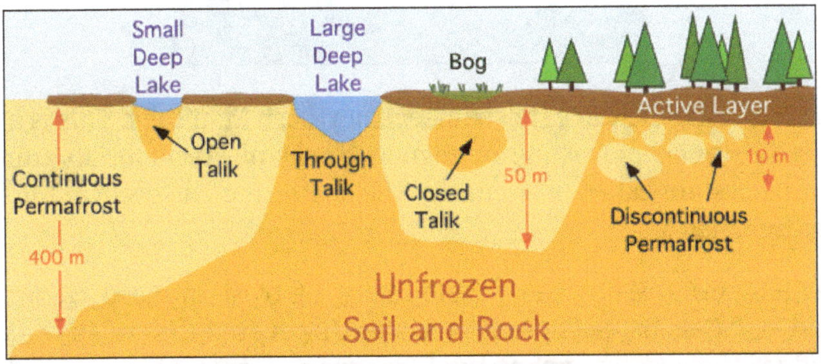

Vertical cross section of the transition zone between continuous and discontinuous permafrost. The graphic also shows the various types of talik or unfrozen ground. Open talik is an area of unfrozen ground that is open to the ground surface but otherwise enclosed in permafrost. Through talik is unfrozen ground that is exposed to the ground surface and to a larger mass of unfrozen ground beneath it. Unfrozen ground encased in permafrost is known as a closed talik.

Five different types of permafrost have been recognized. Continuous permafrost exists over the landscape as an uninterrupted layer. Zones of permafrost with numerous scattered small thawed areas are called discontinuous permafrost. Discontinuous permafrost in usually exists as an extensive marginal zone at the edge of continuous permafrost. Sporadic permafrost occurs where small islands of permafrost are scattered in generally unfrozen areas. Alpine permafrost is found at higher elevations in areas outside where continuous or discontinuous permafrost is common. Subsea permafrost is frozen ground that exist in the sediments beneath seawater. Large areas of subsea permafrost exist along the northern coastal edge of Russia, Alaska, and parts of northern Canada.

Periglacial Processes: Weathering

Locations that have a periglacial environment are characterized by the presence of large quantities of angular, fractured rock. The angular nature of these deposits suggests that the process responsible for the rock fracturing is the crystallization of water. The quantity of the deposits indicates that the frost weathering process operates over and over again in repeated cycles of freeze-thaw. In fact, repeated thawing allows further fracturing because the liquid water is able to fill newly developed cracks.

Frost-shattered granite bedrock (felsenmeer), northern Manitoba. This is a close-up of frost-shattered bedrock, consisting of angular blocks of Precambrian granite. Extensive areas of blocks are called felsenmeer. The shattering has occurred in permafrost terrain, near treeline.

The critical temperature for the development of frost induced fracturing is between -4 and -15° Celsius. Scientists have also discovered that relative slow rates of cooling produce the greatest quantity of fracturing stress. Measurements for marble and granite indicate that a rate of cooling between -0.1 and -0.5° per hour is optimal.

Periglacial Processes: Ground Ice

Surface soils and sediments in periglacial environment are frequently influenced by a variety of different types of ground ice. Some of these masses of ground ice can be as much as 30 meters across. The most common form of ground ice is pore ice. Pore ice develops in the pore spaces between soil and sediment particles where liquid water can accumulate and freeze. Another common form of ground ice is needle ice. Needle ice consists of groups of narrow ice slivers that are up to several centimeters long. They normally form in moist soils when temperatures drop below freezing overnight. Needle ice plays an active role in loosening soil for erosion and tends to move small rocks upward to the soil surface. On sloped surfaces, needle ice can also enhance soil creep by moving soil particles at right angles to the grade.

Ice wedges are downward narrowing masses of ice that are between 2 to 3 meters wide at the base and extend below the ground surface up to 10 meters. It is believed that they form when a seasonal crack in the ground forms in the winter. During this season, extreme cold temperatures can cause soil contraction. Soil and ice, like other solid materials, contracts as temperatures drop way below 0° Celsius. At first, the crack is several millimeters wide and about a meter deep. When temperatures warm up in the summer, liquid water from the active layer fills the crack.

This water then refreezes because the fracture extends into the sub-zero permafrost. The freezing of the water results in a volumetric expansion of about 9%. This expansion then increases the width and depth of the fracture. At the same time, the soil layers adjacent to the developing ice wedge are pushed outward and upward creating a slight hill at the surface. The processes just described cause the ice wedge to increase in size with each passing year. Each winter new cracks form in the ice wedge because of contraction. In the summer, more liquid water seeps in to the cracks to become frozen and the summer warming results in the surface mound getting bigger. Each cycle of ice addition can be identified as narrow layers called foliations. In arid cold environments, some ice wedges can accumulate quantities of wind-blown sand within their winter cracks. These particular types of ground ice are often given the special name sand wedges.

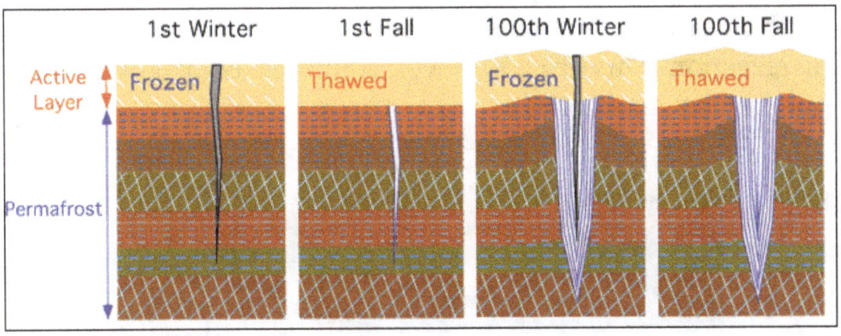

The following series of graphics showing the evolution of ice wedges. The development of an ice wedge begins with cold temperatures in winter forming a crack in the ground surface that is about a meter deep and a few millimeters wide. In the warmer months, melting of the active layer liberates liquid water which flows into the crack. This water then freezes on contact with the permafrost. The cycle just described repeats itself year after year with new cracks forming in the developing ice wedge. The two graphics on the right show a well-developed ice wedge after many cycles of development.

Segregated ice are masses of almost pure ice that grow within permafrost because of liquid water diffusion. Many forms of segregated ice exist as extensive horizontal layers. In this form, they are sometimes called ice lenses. In the development of these features, water migrates under some type of pressure from unfrozen parts of the soil to the ice mass. When the liquid water comes in contact with the segregated ice mass it freezes enlarging the feature. Water diffusion can be caused by temperature and pressure gradients, gravity, and by the movement of groundwater under pressure. Most masses of segregated ice are found just below the active layer. The pressure exerted by overlying sediments controls the growth of segregated ice.

Periglacial Processes: Mass Movement

During the warmer seasons, mass movement can be a common phenomenon in periglacial environments. It usually occurs in four forms: solifluction; gelifluction; frost creepand rockfalls. Solifluction is the slow downslope flow of soil and sediment that is saturated with water. This process can occur on very shallow grades. The common sign of this form of mass movement is the presence solifluction lobes, tongue like semi-mixed surface deposits. In periglacial environments, solifluction is confined to times when temperatures are well above zero and free liquid water is available in the active layer. Solifluction is very common when surface sediments are poorly drained and quite saturated with water. Gelifluction is a form of solifluction where the moving materials slide over a slick permafrost layer.

Frost creep is the slow downslope movement of soil and sediment because of frost heaving and thawing. The process begins with the freezing of the ground surface elevating particles at right angles to the slope. The particles are elevated because cold temperatures cause water in between particles to freeze and expand. In the warm season, thawing causes the ice to convert back to liquid water and the contracting surface drops the particles in elevation. This drop, however, is influenced by gravity causing the particle to move slightly downslope.

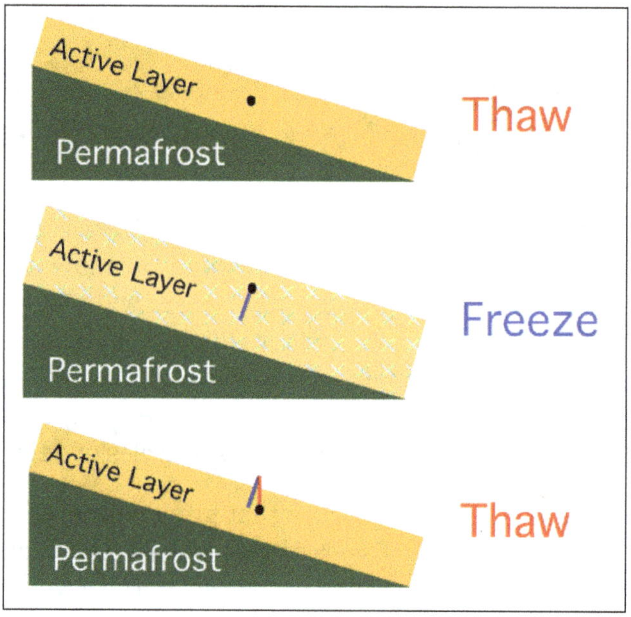

The following series of graphics show how frost creep moves soil particles gradually downslope. The top graphic shows the position of a single particle in a thawed active layer above a zone of permafrost. When cold temperatures occur the active layer freezes. This freezing causes the water between the soil particles to expand making the layer thicker. The expansion processes also moves the illustrated soil particle up. This movement occurs perpendicular to the slope. With the return of warmer temperatures, the active layer contracts as the ice between soil particles melts. However, this contraction is controlled by the force of gravity causing the illustrated soil particle to move forward as it drops.

Exposed rock outcrops are common in periglacial regions of the world. Cold temperatures experienced here do not favor the development of an extensive layer of soil and regolith over bedrock. As a result, exposed rock on slopes is subject to this region's harsh environment. Extreme variations in temperature can cause the fracturing of rock along natural bedding planes and joints leading to rockfalls. This fracturing is mainly due to frost wedging and insolation weathering.

References

- What-are-landforms-and-major-types-of-landforms-on-earth, geology: eartheclipse.com, Re-trieved 10 July, 2020
- Fluvial-landscapes: geography.name, Retrieved 19 June, 2020
- Twidale, C.R., and J.R.V. Romani (2005) Landforms and Geology of Granite Terrains. A.A. Balke-ma Publishers Leiden, The Netherlands. 359 pp. ISBN 978-0415364355
- Glacial-landform, science: britannica.com, Retrieved 07 February, 2020
- Coastal-landform, science: britannica.com, Retrieved 27 August, 2020
- Erosional-landforms-4122800: thoughtco.com, Retrieved 10 March, 2020
- Wang, Zhaoyin; Li, Zhiwei; Xu, Mengzhen; Yu, Guoan (2016-03-30). River Morphodynamics and Stream Ecology of the Qinghai-Tibet Plateau. CRC Press. ISBN 9781315682983

PERMISSIONS

All chapters in this book are published with permission under the Creative Commons Attribution Share Alike License or equivalent. Every chapter published in this book has been scrutinized by our experts. Their significance has been extensively debated. The topics covered herein carry significant information for a comprehensive understanding. They may even be implemented as practical applications or may be referred to as a beginning point for further studies.

We would like to thank the editorial team for lending their expertise to make the book truly unique. They have played a crucial role in the development of this book. Without their invaluable contributions this book wouldn't have been possible. They have made vital efforts to compile up to date information on the varied aspects of this subject to make this book a valuable addition to the collection of many professionals and students.

This book was conceptualized with the vision of imparting up-to-date and integrated information in this field. To ensure the same, a matchless editorial board was set up. Every individual on the board went through rigorous rounds of assessment to prove their worth. After which they invested a large part of their time researching and compiling the most relevant data for our readers.

The editorial board has been involved in producing this book since its inception. They have spent rigorous hours researching and exploring the diverse topics which have resulted in the successful publishing of this book. They have passed on their knowledge of decades through this book. To expedite this challenging task, the publisher supported the team at every step. A small team of assistant editors was also appointed to further simplify the editing procedure and attain best results for the readers.

Apart from the editorial board, the designing team has also invested a significant amount of their time in understanding the subject and creating the most relevant covers. They scrutinized every image to scout for the most suitable representation of the subject and create an appropriate cover for the book.

The publishing team has been an ardent support to the editorial, designing and production team. Their endless efforts to recruit the best for this project, has resulted in the accomplishment of this book. They are a veteran in the field of academics and their pool of knowledge is as vast as their experience in printing. Their expertise and guidance has proved useful at every step. Their uncompromising quality standards have made this book an exceptional effort. Their encouragement from time to time has been an inspiration for everyone.

The publisher and the editorial board hope that this book will prove to be a valuable piece of knowledge for students, practitioners and scholars across the globe.

INDEX

www.ingramcontent.com/pod-product-compliance
Lightning Source LLC
LaVergne TN
LVHW080629210926
848485LV00003B/235